VOLUME EIGHTY NINE

Advances in
PARASITOLOGY

SERIES EDITOR

D. ROLLINSON
Life Sciences Department
The Natural History Museum,
London, UK
d.rollinson@nhm.ac.uk

J. R. STOTHARD
Department of Parasitology
Liverpool School of Tropical
Medicine Liverpool, UK
russell.stothard@lstmed.ac.uk

EDITORIAL BOARD

T. J. C. ANDERSON
Department of Genetics, Texas
Biomedical Research Institute,
San Antonio, TX, USA

M. G. BASÁÑEZ
Professor of Neglected Tropical
Diseases, Department of Infectious
Disease Epidemiology, Faculty of
Medicine (St Mary's Campus),
Imperial College London,
London, UK

S. BROOKER
Wellcome Trust Research Fellow
and Professor, London School of
Hygiene and Tropical Medicine,
Faculty of Infectious and Tropical,
Diseases, London, UK

R. B. GASSER
Department of Veterinary Science,
The University of Melbourne,
Parkville, Victoria, Australia

N. HALL
School of Biological Sciences,
Biosciences Building, University of
Liverpool, Liverpool, UK

J. KEISER
Head, Helminth Drug
Development Unit, Department
of Medical Parasitology and
Infection Biology, Swiss Tropical
and Public Health Institute, Basel,
Switzerland

R. C. OLIVEIRA
Centro de Pesquisas Rene Rachou/
CPqRR - A FIOCRUZ em Minas
Gerais, Rene Rachou Research
Center/CPqRR - The Oswaldo Cruz
Foundation in the State of Minas
Gerais-Brazil, Brazil

R. E. SINDEN
Immunology and Infection
Section, Department of Biological
Sciences, Sir Alexander Fleming
Building, Imperial College of
Science, Technology and
Medicine, London, UK

D. L. SMITH
Johns Hopkins Malaria Research
Institute & Department of
Epidemiology, Johns Hopkins
Bloomberg School of Public Health,
Baltimore, MD, USA

R. C. A. THOMPSON
Head, WHO Collaborating Centre
for the Molecular Epidemiology
of Parasitic Infections, Principal
Investigator, Environmental
Biotechnology CRC (EBCRC), School
of Veterinary and Biomedical
Sciences, Murdoch University,
Murdoch, WA, Australia

X.-N. ZHOU
Professor, Director, National
Institute of Parasitic Diseases,
Chinese Center for Disease Control
and Prevention, Shanghai, People's
Republic of China

VOLUME EIGHTY NINE

Advances in
PARASITOLOGY

Edited by

D. ROLLINSON
Life Sciences Department
The Natural History Museum
London, UK

J. R. STOTHARD
Department of Parasitology
Liverpool School of Tropical Medicine
Liverpool, UK

AMSTERDAM • BOSTON • HEIDELBERG • LONDON
NEW YORK • OXFORD • PARIS • SAN DIEGO
SAN FRANCISCO • SINGAPORE • SYDNEY • TOKYO
Academic Press is an imprint of Elsevier

Academic Press is an imprint of Elsevier
125 London Wall, London, EC2Y 5AS, UK
The Boulevard, Langford Lane, Kidlington, Oxford OX5 1GB, UK
225 Wyman Street, Waltham, MA 02451, USA
525 B Street, Suite 1800, San Diego, CA 92101-4495, USA

First edition 2015

Copyright © 2015 Elsevier Ltd. All rights reserved.

No part of this publication may be reproduced or transmitted in any form or by any means, electronic or mechanical, including photocopying, recording, or any information storage and retrieval system, without permission in writing from the publisher. Details on how to seek permission, further information about the Publisher's permissions policies and our arrangements with organizations such as the Copyright Clearance Center and the Copyright Licensing Agency, can be found at our website: www.elsevier.com/permissions.

This book and the individual contributions contained in it are protected under copyright by the Publisher (other than as may be noted herein).

Notices
Knowledge and best practice in this field are constantly changing. As new research and experience broaden our understanding, changes in research methods, professional practices, or medical treatment may become necessary.

Practitioners and researchers must always rely on their own experience and knowledge in evaluating and using any information, methods, compounds, or experiments described herein. In using such information or methods they should be mindful of their own safety and the safety of others, including parties for whom they have a professional responsibility.

To the fullest extent of the law, neither the Publisher nor the authors, contributors, or editors, assume any liability for any injury and/or damage to persons or property as a matter of products liability, negligence or otherwise, or from any use or operation of any methods, products, instructions, or ideas contained in the material herein.

ISBN: 978-0-12-803301-2
ISSN: 0065-308X

For information on all Academic Press publications
visit our website at store.elsevier.com

CONTENTS

Contributors — vii

1. Ecology of Free-Living Metacercariae (Trematoda) — 1
Neil J. Morley

1. Introduction — 2
2. Transport Hosts and Metacercariae: Some Basic Ecological Concepts — 4
3. Presettlement Phase — 9
4. Settlement Phase — 20
5. Aberrant Free-Living Existence — 52
6. Metacercarial Biology — 54
7. Pollution and Free-Living Metacercariae — 60
8. Concluding Remarks — 62
References — 64

2. Cross-Border Malaria: A Major Obstacle for Malaria Elimination — 79
Kinley Wangdi, Michelle L. Gatton, Gerard C. Kelly, Archie CA. Clements

1. Introduction — 80
2. Patterns of Movement — 82
3. Epidemiological Drivers of Malaria in Border Areas — 89
4. Way Forward — 92
5. Conclusion — 97
Contributors — 98
References — 99

3. Development of Malaria Transmission-Blocking Vaccines: From Concept to Product — 109
Yimin Wu, Robert E. Sinden, Thomas S. Churcher, Takafumi Tsuboi, Vidadi Yusibov

1. Introduction — 110
2. Targets for TBVs — 113
3. Vaccine Development Efforts and Status — 121

4. Protective Correlates and Surrogate Assays for Efficacy	131
Acknowledgements	138
References	138

Index   *153*

Contents of Volumes in This Series   *159*

CONTRIBUTORS

Thomas S. Churcher
MRC Centre for Outbreak Analysis and Modelling, Department of Infectious Disease Epidemiology, School of Public Health, Imperial College London, London, UK

Archie CA. Clements
The Australian National University, Research School of Population Health, College of Medicine, Biology and Environment, Canberra, ACT, Australia

Michelle L. Gatton
Queensland University of Technology, School of Public Health & Social Work, Brisbane, Qld, Australia

Gerard C. Kelly
The Australian National University, Research School of Population Health, College of Medicine, Biology and Environment, Canberra, ACT, Australia

Neil J. Morley
School of Biological Sciences, Royal Holloway, University of London, Egham, Surrey, UK

Robert E. Sinden
The Jenner Institute, Oxford, UK

Takafumi Tsuboi
Division of Malaria Research, Ehime University, Matsuyama, Ehime, Japan

Kinley Wangdi
The Australian National University, Research School of Population Health, College of Medicine, Biology and Environment, Canberra, ACT, Australia; Phuentsholing General Hospital, Phuentsholing, Bhutan

Yimin Wu*
Laboratory of Malaria Immunology and Vaccinology, National Institute of Allergy and Infectious Diseases, Rockville, MD, USA

Vidadi Yusibov
Fraunhofer USA Center for Molecular Biotechnology, Newark, DE, USA

---

*Current affiliation: PATH-Malaria Vaccine Initiative, Washington DC, USA

CHAPTER ONE

Ecology of Free-Living Metacercariae (Trematoda)

Neil J. Morley
School of Biological Sciences, Royal Holloway, University of London, Egham, Surrey, UK
E-mail: n.morley@rhul.ac.uk

Contents

1. Introduction	2
2. Transport Hosts and Metacercariae: Some Basic Ecological Concepts	4
3. Presettlement Phase	9
3.1 Cercarial morphology	9
3.2 Cercarial emergence	9
3.2.1 Patterns of emergence	9
3.2.2 Abiotic effects: temperature	10
3.2.3 Abiotic effects: other factors	11
3.2.4 Biotic effects	12
3.3 Dispersive phase	13
3.3.1 Cercarial swimming	13
3.3.2 Orientation behaviour	13
3.3.3 Duration of dispersive phase	16
3.3.4 Himasthla rhigendana as a model species for dispersal	19
4. Settlement Phase	20
4.1 Settlement and encystment	20
4.2 Gregarious and substrate-associated settlement behaviour	21
4.2.1 Cyst aggregations	23
4.2.2 Cyst associations	23
4.3 Settlement and encystment cues	25
4.3.1 Chemical cues	25
4.3.2 Physical cues	27
4.3.3 Encystment cues for free-floating metacercariae	29
4.4 Metacercarial cyst morphology	29
4.4.1 Morphology of fixed metacercariae	29
4.4.2 Morphology of floating metacercariae	31
4.5 Transport hosts	31
4.5.1 Host selection	31
4.5.2 Molluscan transport hosts	33
4.5.3 Crustacean transport hosts	38
4.5.4 Plant transport hosts	40
4.5.5 Miscellaneous transport hosts	44

4.6 Free-floating metacercariae	46
4.6.1 Principal free-floating metacercariae	46
4.6.2 Secondary free-floating metacercariae	48
5. Aberrant Free-Living Existence	52
6. Metacercarial Biology	54
6.1 Metacercarial viability	54
6.1.1 Abiotic effects: temperature	54
6.1.2 Abiotic effects: other factors	56
6.2 Metacercarial infectivity	57
6.3 Metacercarial excystment	58
7. Pollution and Free-Living Metacercariae	60
8. Concluding Remarks	62
References	64

Abstract

The presence of trematodes with a free-living metacercarial stage is a common feature of most habitats and includes important species such as *Fasciola hepatica*, *Parorchis acanthus* and *Zygocotyle lunata*. These trematodes encyst on the surface of an animal or plant that can act as a transport host, which form the diet of the target definitive host. Although these species are often considered individually, they display common characteristics in their free-living biology indicating a shared transmission strategy, yet in comparison to species with penetrative cercariae this aspect of their life cycles remains much overlooked. This review integrates the diverse data and presents a novel synthesis of free-living metacercariae using epibiosis as the basis of a new framework to describe the relationship between transport hosts and parasites. All aspects of their biology during the period that they are metabolically independent of a host are considered, from cercarial emergence to metacercarial excystment.

1. INTRODUCTION

The metacercariae is a larval stage that is common to many species of trematodes. It represents a phase where the potential to continue the life cycle and the ability to infect definitive hosts is retained and extended over a relatively long period. The morphology and physiology of this stage is thus organized to this end and is often regarded as a 'resting' stage, although this should not imply a period of total physiological inactivity, a cessation of metabolic processes, nor the absence of development (Erasmus, 1972). Infectivity of metacercariae is passive, occurring only when the target host ingests the parasite. Distribution is therefore closely linked to the food chain and the feeding pattern of the definitive host. Dönges (1969) determined that trematodes could be separated into three groups on the basis of their metacercarial biology: (1) species whose metacercariae encyst freely in the open; (2) species whose metacercariae penetrate and encyst within a second intermediate host and (3) metacercarial species which undergo growth and metamorphosis after penetrating into the second intermediate host. The spatial and temporal distribution of metacercarial populations is generally dependent on the prior life history stage, the free-living cercariae. Although much general attention has focused on the attributes of group's 2 and 3 (e.g. Chubb (1979) and Keeler and Huffman (2009)); consideration of free-living metacercariae as a group remain surprisingly neglected, with most overviews focusing on specific parasite species such as *Fasciola hepatica* (Dalton, 1999).

Free-living metacercariae usually encyst on the surface of an animal or plant that can act as a transport host, that is, one where there is physical contact between host and parasite but no physiological interaction (*sensu* Odening, 1976), which the definitive host will subsequently feed upon. As trematodes are capable of demonstrating complex and subtle variations in life cycles, it is therefore important to determine what exactly constitutes a free-living existence. For the purposes of this article I designate 'free-living metacercariae' to include those species that neither reside within host tissue nor reach a settlement site on the transport host by penetrating through tissue. They also include those species that adopt a free-floating lifestyle, unattached to a transport host, either as a principal or secondary aspect of their free-living existence. I use the term 'free-living metacercariae' freely throughout as a general definition of this group of trematodes regardless of the specific life stage under consideration. Metacercariae of penetrative species that may become free-living due to atypical mechanisms I designate 'aberrant'. Although such forms are not the main focus of this work I briefly discuss the different types occurring in the environment as they represent a mode of free life that, although unrepresentative, may still potential contribute a viable route of infection and thus such a free-living existence may be important for the ecology of those species.

The vast majority of focus on free-living stages of trematodes has been centred on species of cercariae that actively seek out and penetrate a target host. In comparison, free-living metacercariae remain much overlooked and misunderstood. They are often unnecessarily relegated to a minor status when considering the ecology of trematode transmission, with a perception that their distribution within a habitat involves only a crude level of behaviour. Yet, for these species this stage represents the vast majority of the free-living life, capable of stretching into many months. The strategies employed by this group to increase transmission success are sophisticated and the equal of those that penetrate into the tissues of target hosts. Encystment site selection by free-living metacercariae represents a critical behaviour during settlement because the right choice of transport host and the precise location on it can have a significant impact on long-term viability and successful transmission. The spatial and temporal distribution of these metacercariae in the environment therefore plays a key role in the epidemiology of the parasite in the vertebrate definitive host.

In an extension of the long-standing idea that cercariae can be viewed as being akin to meroplankton (see for example Morley (2012)), the ecology and formation of free-living metacercarial populations can be seen as analogous to the distribution and settlement of marine invertebrate larvae, particularly epibionts that attach and grow on the living surface of another organism. Thus, many aspects of the well-known behaviour and many of the settlement cues used by these larvae (e.g. McEdwards (1995)), may be applicable to free-living metacercariae and could provide useful pointers to the mechanisms that influence the encystment processes of such trematodes. Similarly, the relationship epibionts have with the basibiont (substrate organism) may provide an insight into potential interactions between free-living metacercariae and transport hosts.

The establishment of free-living metacercarial populations that are attached to a transport host can be viewed as incorporating two phases — an initial presettlement phase where the cercariae emerge, disperse and encyst on a chosen host and a final settlement phase where the established metacercariae passively wait to be ingested by the target definitive host. Nevertheless, metacercarial species that are principally free-floating adopt a much wider range of strategies to form their free-living populations that are in general more poorly documented.

This chapter is concerned only with the period when these species are metabolically independent of a host, relying solely on their glycogen store for energy. It therefore encompasses all aspects in the formation and duration of free-living metacercarial populations from cercarial emergence to the excystment of the parasite upon ingestion by the definitive host.

2. TRANSPORT HOSTS AND METACERCARIAE: SOME BASIC ECOLOGICAL CONCEPTS

From a general assessment of the relevant scientific literature it is apparent that there is a degree of underestimation and confusion regarding the exact nature of the relationship between transport host and free-living metacercariae. It was therefore considered worthwhile to set out a few basic concepts in order to place these associations in the context of more traditional host–parasite interactions. Although the definition of transport hosts as a theoretical model has been much debated (Sprent, 1963; Odening, 1976; Macko, 1980), many of the fundamental practical aspects of the ecological relationship between free-living metacercariae and chosen transport host remain sadly neglected, simply because they interact on a much more limited basis compared to the intricate ecophysiological relationships present in more standard host–parasite associations. Nevertheless, the vast amount of literature on epibiosis provides a relevant basic conceptual framework (e.g. Wahl (1989)), for evaluating likely key aspects of their interactions. However, this can only represent potential scenarios and should not be interpreted as indicating any degree of definitive evidence for these kinds of interactions between metacercariae and transport hosts.

To begin with it will be useful to address weaknesses of a couple of common basic terms used in relation to free-living metacercarial ecology. Firstly, transport hosts are often in the scientific literature simply referred to as 'substrate'. Although this is technically correct, it does give the false impression that the host has the same level of interactions with the metacercariae as an inert surface would. Evidence from epibiosis studies would suggest that this is unlikely to be the case and therefore 'transport host' may be a better general term, with 'substrate' confined to defining the exact location on the host where encystment takes place. Secondly, many studies describe species as encysting 'on any hard surface' (e.g. Yamaguti (1975)). Such statements are unhelpful as observations usually took place under artificial laboratory conditions in glass containers or on glass slides, presenting the cercariae with an overload of cues from hard surfaces, potentially masking other settlement prompts. It seems unlikely that under natural conditions cercariae would adopt such indiscriminate behaviour, particularly where many habitats, such as coastal rocky shores, have bed rock which represents an abundance of hard surfaces that cannot possibly aid the transport of the parasite to the target host. Use of this term should therefore be avoided.

Both metacercariae and transport host will potentially benefit or be hindered by their relationship. For metacercariae, determining the most optimum form of transport host must be a key objective. In theory, because the transmission of metacercariae is dependent on trophic interactions any organism that regularly forms the diet of the target definitive host may have the potential to act as a transport host. Laboratory studies have generally demonstrated that many species of free-living metacercariae will encyst on a wide range of organisms. However, field studies are rare and it therefore remains largely unknown if such experimental host choices would be replicated under natural conditions, although the available evidence would suggest that some differences occur (see Section 4.5.1). Host choice is likely to be determined by a number of factors. Settlement by metacercariae will probably occur in hydrodynamically favourable sites, either within habitats or upon individual transport hosts, presumably where turbulence is reduced and potential transport hosts remain visible to predators/grazers.

A significant problem for free-living metacercariae is the unstable character of the living substratum of the transport host, with both natural mortality and physical disturbance of the transport host potentially interfering with the parasites transmission success. Further instability is caused by morphological changes of the transport host during its life cycle. In a similar manner to those changes encountered by epibionts (Wahl, 1989), these variations can include host growth and shrinking, fission and fusion, shedding of fruiting bodies and leaves, and moulting. The longer a metacercariae remains attached to a transport host the greater the risk morphological changes may pose to the parasites viability.

Similarly, fluctuations in the physiological activity of the transport host may also induce instability. Production and exudation of metabolites such as waste, nutrients and toxins may change both temporally with season, predation pressure, developmental stage and biological cycles and spatially according to habitat and different locations on the same host (Wahl, 1989). Although metacercarial cysts are considered to show a great tolerance to a range of abiotic environmental conditions (see Section 6.1), their responses to biotic conditions associated with the transport host remain unknown and their viability may be dependent on phases when, or on host sites, where the composition and quantities of exudate are not considered harmful.

A final risk factor for metacercariae is the dangers of physiological stress due to potential drastic environmental changes associated with the transport host migrations or shifts in habitat stability. For example, transport hosts in marine intertidal zones will experience daily extremes in temperature, salinity and humidity, while the growth of plant transport hosts may shift metacercariae encysted upon them from favourable to unfavourable climatic conditions over time.

For the transport hosts themselves colonization by free-living metacercariae may indirectly affect their own ecological and physiological homeostasis, although such changes are likely to depend on the hosts relative size compared to the parasite, smaller hosts being more vulnerable, and the intensity of the metacercarial population. A covering of metacercariae may potentially have a protective role, possibly slowing down the rate of desiccation in air-exposed hosts or by providing a 'camouflage' through their own surface properties that may interfere with the cues used by nontarget host predators to identify prey.

More often, free-living metacercariae will likely have detrimental effects on the transport host. Colonization by parasites may cause an increase in weight and consequently reduce buoyancy, while rigid cysts may reduce the elasticity of the settlement site on the host hindering motion and flexibility, thereby increasing the likelihood of breakage in highly turbulent environments. Metacercariae may cause shading of plant surfaces, reducing photosynthesis, which under particularly heavy parasite burdens may induce a negative energy budget resulting in the ultimate mortality of the host. Similarly, transcutaneous uptake or excretion of nutrient salts, dissolved organic matter, ions and/or gases by transport hosts may be affected, although any impact will be dependent on the specific organism's biology and the settlement location of the parasite.

It is clear that free-living metacercariae have the potential to be unfavourable for the transport hosts ecology and functional biology. Many organisms, in order to prevent colonization by epibionts have developed a number of strategies to protect themselves (Wahl, 1989; Wahl et al., 1998; Krug, 2006), which may also protect against free-living metacercariae. Three classes of adaptations are considered to have evolved: tolerance, avoidance and defence (Wahl, 1989). Many species, particularly sedentary ones, which secrete mineral outer shells are capable of tolerating extensive colonization of these surfaces. Nevertheless, an important prerequisite for this tolerance is a prevailing indifference to increased friction and weight and a physiological inactive outer surface. However, even in heavily colonized individuals the body's orifices (shell borders, tubes and siphons) and external sense organs are maintained clean (Wahl, 1989).

Avoidance of colonization may include movement in space, time or dissimulation. Accelerated growth or reproduction may produce tissue or offspring at a rate higher than the colonization rate; this is particularly relevant for plant growth which may allow for the continued maintenance of a photosythetically active zone (e.g. Sand-Jensen (1977) and Bultlauist and Woelkerling (1983)). Similarly, the migrations of species into biologically less vulnerable habitats or optical and chemical camouflage may also be considered avoidance.

Defensive mechanisms can include mechanical, physical and chemical elements. Mechanical aspects that may impede colonization include special surface structures such as spicules, intense surface production of mucus, periodic shedding of the cuticula or epidermis, scale-casting, friction between the body surface of burrowing or fast-swimming species and the surrounding sediment or water, and the active removal of colonizers by the scraping of the surface with specialized appendages (Dyryndd, 1986; Wahl, 1989; Wahl et al., 1998). Nevertheless, the efficiency of such mechanical mechanisms depends on many factors such as the proportion of the surface cleaned at any given time, the size and intensity of colonies, and the regularity of surface disturbance by the hosts' mechanical defences. The properties of the physical surface of transport hosts, in particular low-energy surfaces where there is a minimum of free ions in the outermost layer, such as found in certain waxes or nonpolar hydrocarbons can also reduce the amount of colonization. However, such surfaces primarily impede the adhesion rather than initial settlement of organisms, and therefore a high relative water velocity is required for this kind of mechanisms to be efficient (Wahl, 1989). Chemical surface defences appear to be widespread in aquatic animals and include extreme pH values or toxic secondary metabolites exuded or surface bound to host species which are considered detrimental to colonization.

Defence mechanisms are frequently combined to form a multifaceted protection against epibionts. For any given species the resulting complex of adaptation may vary with season, latitude, habitat, individual age, physiological condition, biological cycles and tissue type (Wahl, 1989; Wahl et al., 1998). For free-living metacercariae, as for epibionts, this may result in spatial and temporal variability's in their distribution and consequently influence the chances of transmission to the definitive host.

The framework described in this section presents a wide range of potential ecological scenarios that may influence the establishment of free-living metacercariae. Some will be more likely than others to affect parasite occurrence, while others will have relevance only for certain species combinations or habitats. Nevertheless, it is important to understand the breadth of variables involved, although only a few may have so far been documented in relation to these parasites. Examples from specific metacercariae—transport host associations which demonstrate some of these concepts can be found in the following sections.

3. PRESETTLEMENT PHASE
3.1 Cercarial morphology

In general, the structure of these cercarial species follows the same basic morphotype, reflecting their shared ecology and behaviour rather than phylogenetic relationships. They have a large oval or elongated body with a tail that is often of the same length or larger (Galaktionov and Dobrovolskij, 2003), and possess a larger body volume than those species penetrating vertebrates (Koehler et al., 2012). As these species do not grow until they reach the definitive host, the size of the cercariae probably influences the size and future fitness of the adult creating a strong selection pressure for a large cercarial body (Koehler et al., 2012). Many species, notably the Notocotylids, have dark pigmented bodies due to melanin (e.g. Wunder (1932)), which is considered to provide the cercariae with some degree of protection against UV light and desiccation (Nedakal, 1960) and may also provide the same function for encysted metacercariae.

Typically, numerous cytogenous glands are found in the body with many species possessing pigmented eyespots. Specialized penetration structures, such as a stylet or penetration glands, are absent. Species may have either one (oral) or two (oral and ventral) suckers with species having only one sucker often developing other specialized ventral attachment structures such as locomotory appendages on either side of the base of the tale, as found in the Notocotylidae (Galaktionov and Dobrovolskij, 2003). In some species, such as Philophthalmidae, the terminal part of the tail has a sucker-like structure which, with associated glandular secretions, allows the cercariae to adhere to surfaces and stand up on the tip of its tail in order to perform 'searching' behaviour patterns while remaining in the same place, designated as 'ambuscade' (Galaktionov and Dobrovolskij, 2003).

3.2 Cercarial emergence
3.2.1 Patterns of emergence

For free-living metacercarial species relatively few cercariae emerge from snails on a daily basis, numbers typically ranging from less than 10 up to the high hundreds from any individual mollusc, and this production level is significantly lower than that found for cercarial species that infect vertebrate hosts (Thieltges et al., 2008).

The emergence of cercariae often follows distinct daily patterns resulting in surges in parasite numbers during key time periods that are probably synchronized with the targets' host diurnal behavioural patterns in order to ensure an increased chance of infection. However, there are few general applicable circadian or infradian patterns of emergence between different free-living metacercarial species. Those that utilize plant transport hosts such as *Fasciola hepatica* and *Fasciola gigantica* tend to emerge in very low daily numbers with little evidence of circadian rhythm, although the latter demonstrates a preference for emergence during the night. Arrhythmic emergence in these species is unsurprising as the plant hosts do not have active cycles and therefore always remains available to the searching cercariae making any synchronization unnecessary (Bouix-Busson et al., 1985). *Fasciola hepatica* emergence has been found to be periodic, with releases occurring in waves separated by an interwave period of 6–8 days. Up to four or five emergence periods may occur before the death of the snail host (Dreyfuss and Rondelaud, 1994; Vignoles et al., 2006). However, utilizing a plant transport host does not always result in an absence of emergence patterns. Studies on the amphistome *Cotylophoron cotylorum* demonstrate a diurnal pattern but have generated contradictory results either showing a peak between 11 am and 2 pm (Bennett, 1936; Varma, 1961) or in the early hours of the morning, before 9 am (Krull, 1934).

In contrast, Notocotylids, parasitizing mainly molluscan and crustacean transport hosts, typically emerge daily in numbers of the low hundreds range during the late morning/early afternoon period (Pike, 1969; Harris, 1986; Besprozvanykh, 2010). However, the marine species *Parorchis acanthus*, utilizing a similar range of transport hosts, emerges at night or the early morning (Stunkard and Shaw, 1931) as do a small number of other Philophthamid and Notocotylid species (West, 1961; Smith and Hickman, 1983), possibly to take advantage of highly mobile crustacean hosts when they are at their least active. Other species targeting plant transport hosts may also preferentially emerge at night (Pike, 1968; Martin, 1973), however the majority of free-living metacercariae emerge during the day, either in the morning (Dutt and Srivastava, 1972; Shameem and Madhavi, 1991; Jousson and Bartoli, 1999; Diaz et al., 2009) or in the afternoon (Krull and Price, 1932; Murrell, 1965; Stunkard, 1967a) suggesting light is a prominent mechanism controlling emergence regardless of their target transport host.

3.2.2 Abiotic effects: temperature

Other factors in addition to light may also be influential, and experimental manipulation of the environmental conditions necessary for cercarial emergence has been extensively investigated. However, laboratory studies on cercarial emergence must be treated with some caution as this process is highly sensitive to changes in abiotic conditions, artificially stimulating the release of cercariae producing false-positive results, and thus careful acclimation is required to achieve reliable data (Morley and Lewis, 2013). Temperature is a key variable that may influence emergence and has been extensively studied in free-living metacercariae (see Morley and Lewis (2013)). Almost all cercarial species regardless of their life histories follow the same pattern of responses to temperature. Emergence rates of increasing temperature are different between mid- and low-latitude species and can be divided into three broad ranges — a suboptimal low-temperature range where relatively few cercariae emerge (10–20 °C for mid-latitude species and 15–25 °C for low-latitude species), an optimal-temperature range where emergence achieves its maximum values (15–25 °C for mid-latitude species and 20–30 °C for low-latitude species) and a high-temperature range where emergence declines (20–30 °C for mid-latitude species and 25–35 °C for low-latitude species) (Morley and Lewis, 2013). Within the low-temperature range a minimum emergence temperature threshold (METT) exists for each species which is a temperature point where emergence rates decline to almost zero. As temperature rises above this point, the numbers of emerging cercariae rapidly increases. Species such as *Fasciola hepatica* that can be found in both mid- and low-latitude habitats, demonstrate different emergence thermodynamics dependent on the temperature ranges experienced in their individual latitudes that they have adapted to (Morley and Lewis, 2013).

3.2.3 Abiotic effects: other factors

Salinity is also important for the emergence of marine cercariae, and species-specific responses to changes in this parameter are apparent. *Parorchis acanthus* demonstrates a decrease in the numbers of cercariae emerging when salinity is both increased or decreased (Rees, 1948). In contrast, *Philiphthalmus* sp. showed an increased level of emergence with a decline in salinity but remained unchanged when levels increased over a short-term period (Koprivnikar and Poulin, 2009). However long-term exposure of 12 weeks duration resulted in this species having a reduced level of emergence when salinity was lowered. Yet, over the duration of the experiment there was great variability in the sensitivity to lowered salinity from 1 week to another (Lei and Poulin, 2011). These responses may be associated with the individual host snail's responses to salinity as its physiological status has a controlling influence on cercarial emergence.

The water level may also influence the emergence of these species. *Philophthalmus* sp. demonstrated greater emergence when the snail host was completely submerged, although emergence still occurred from partially exposed snails (Koprivnikar and Poulin, 2009). Similarly, *Parorchis acanthus* emergence was also enhanced by the presence of water, but cercariae were also released in low numbers under only 'damp conditions' where snails placed in dry dishes were moistened with an aerosol spray prior to the start of the experiment, simulating a low tide in the lower eulittoral. Released cercariae were also capable of successful encystment under these conditions, suggesting a parasite transmission window that was independent of tidal cycles (Prinz et al., 2011).

For freshwater species, changes in pH may influence emergence. Varma (1961) found that *C. cotylophorum* retained maximum levels of emergence at a pH of between 6.5 and 9.5. A pH both above and below this range resulted in gradually fewer cercariae emerging with no emergence taking place at pH below 3 or above 11. Similarly, the emergence of *Fasciola hepatica* was unaffected by a pH value within the range of 5.5–8.5 (Kendall and McCullough, 1951). In addition, an increase in CO_2 levels slowed, but did not inhibit the emergence of this species, while oxygen depletion had no apparent effect (Kendall and McCullough, 1951).

3.2.4 Biotic effects

Under certain circumstances encystment without emergence from the first intermediate host can occur (Sonsino, 1892; Wesenberg-Lund, 1934; Vareille-Morel et al., 1993a; Dreyfuss et al., 1995, 2009), and may potentially be a common occurrence for free-living metacercariae. Vareille-Morel et al. (1993a) studied the formation of these cysts in detail for *Fasciola hepatica*. It would appear such internal cysts are formed immediately prior to, or following, the infected snails death and result from mature cercariae within the host tissues reacting to pathological changes associated with this event. Formation of cysts was most common in snails that had harboured infections for at least 70 days before dying with the majority of snails containing only one or two fully formed cysts. It is possible that other factors may also result in the formation of these internal cysts. Echinostome metacercariae will often develop similar infections within the first intermediate host in response to drastic changes in abiotic conditions, such as temperature extremes and pollution (Wesenberg-Lund,

1934; Morley et al., 2004), and free-living metacercariae may potentially respond in a similar manner.

3.3 Dispersive phase

The behaviour of cercariae during the planktonic dispersive phase is important in ensuring that parasites are widely distributed on transport hosts within a habitat to increase the chances of transmission to the definitive host. However, it remains relatively poorly understood on a large scale, and the majority of information is derived from a handful of studies.

3.3.1 Cercarial swimming

During this presettlement phase, swimming is the main mode of dispersal with cercariae needing to control both the direction and speed of locomotion. They propel themselves by contractions of muscles in the tail and, in general, demonstrate a 'continuous swimming' behaviour that ensures they are well adapted to scan exhaustively for potential habitats (Haas, 1994), but the rate of the swimming stroke is generally slower (tail oscillations of 3–20 Hz) than other kinds of cercariae that penetrate target hosts (Rees, 1971; Coil, 1984; Bennett, 2001). For *Fasciola hepatica* at least, the lateral beating of the tail oscillations and the body during swimming ensure a minimum movement of the ventral sucker. In combination with the bulbous sensory endings associated with this organ such swimming adaptations may aid contact detection and initial attachment to surfaces as precursors for encystment (Bennett, 2001).

Lecithotrophic larval swimming speed changes as a function of temperature and also salinity. Changes are always positively thermokinetic, demonstrating increased speed or frequency of locomotory activity. Such effects are partly due to changes in metabolic rates, but may also be associated with thermal differences in viscosity of water that can influence the physiological demands of swimming (Young, 1995). Although nothing is known about the cercarial responses of free-living metacercariae to these parameters, activity of other kinds of cercariae demonstrate an increased level of activity at higher temperatures that can only be maintained for shorter durations, with marked interspecies differences being apparent (Rea and Irwin, 1995; Koprivnikar et al., 2010).

3.3.2 Orientation behaviour

Orientation behaviour is an important component in the distribution of all kinds of cercariae, influencing the spatial position within a habitat and ultimately guiding the parasite into the 'host space' (Combes et al., 1994). There are a wide range of cercarial population models that reflect the distribution of different kinds of targeted hosts (Morley, 2012). It has been found that the physical stimuli that marine invertebrate lecithotrophic larvae may respond to include scalars (salinity, dissolved oxygen, temperature and pressure), which can initiate a kinetic response, and vectors (light, gravity and water flow), which can induce taxis (Young, 1995). For species of free-living metacercariae there is no information on the effects of scalar stimulus to their cercariae and only limited data on vector effects.

3.3.2.1 Responses to light and gravity

Light and gravity are the two main vectors that marine invertebrate larvae orientate towards, and the responses of many species change as they age from positive to negative or visa versa (Young, 1995). The available evidence would suggest that free-living metacercariae also predominantly respond to these vectors; however, age-related changes, demonstrated in other kinds of cercariae such as host-penetrating echinostomes (McCarthy, 1999; Platt and Dowd, 2012), have yet to be quantitatively studied in this group. Nevertheless, a small number of studies suggest that age-related changes in orientation may potentially occur, being described as demonstrating both positive phototactic and geotactic behaviour (Burns, 1961; Macy and Bell, 1968).

In general, positive phototaxis has been shown to occur in the cercariae of a wide range of species (e.g. Bennett (1936), Probert (1965), Murrell (1965), Simon-Vicente et al. (1985) and Shameem and Madhavi (1991)) suggesting it is a common response in this group. Other species accumulate on the bottom of experimental dishes indicating either or both a positive geotactic and negative phototactic response (West, 1961; Erkina, 1954; Hunter, 1967; Al-Jahdali and El-Said Hassanine, 2012). More complex reactions have also been documented with certain species of psilostome cercariae showing positive phototaxis, but not encysting until the light intensity has fallen to low levels (Pike, 1968). Similarly, *Fasciola hepatica* cercariae demonstrated both negative phototropism and geotropism, the parasite being repelled by strong light but still rising to the water surface where they encyst on the underside of leaves (Pecheur, 1967), while *Notocotylus ralli* demonstrates mainly positive geotaxis but will prolong the dispersive phase in low light levels, encystment only being stimulated by the onset of high light intensity (Dönges, 1962).

3.3.2.2 Responses to colour

Although phototactic responses to shaded or well-illuminated surfaces may strongly influence settlement choice, a substantial amount of information is also available on the role of colour. The visual attribute of objects that results from the light they emit, transmit or reflect is the usual definition of colour, and larval reactions to colour may be due to complex factors associated with the quantities of radiant energy absorbed or reflected (Satheesh and Wesley, 2010). With one exception, only the colour preferences of those free-living metacercariae that utilize plant transport hosts have so far been investigated. Amphistomes, including *Paramphistomum cervi*, *P. ichikawai*, *Ceylonocotyle streptocoelium* and *C. calicophorum*, are all attracted to yellow light or painted surfaces (Durie, 1955, 1956; Burgu, 1982), while *P. epiclitum* demonstrates an almost equal preference to both green and yellow polythene strips (Varma and Prasad, 1998). However, *Fasciola hepatica* has been reported to have more contradictory responses to colours. No attraction to either yellow light or any coloured surface has been shown by Durie (1955) and Pecheur (1967) respectively, while Jimenez-Albarron and Guevara-Pozo (1980) found a slight preference for a glass surface externally covered with yellow cellophane compared to surfaces covered in green, red, blue or colourless.

In contrast, a more detailed study by Tripathi et al. (2014) on *Fasciola gigantica* cercarial attraction towards different wavelengths of visible light found that it was dependent on both parasite age and light intensity. Exposure to 150 lux of red light at 650 nm caused the maximum level of attraction, which increased progressively as the cercariae aged from 15 to 60 min postemergence. This appears to be associated with higher actylecholinesterase and cytochrome oxidase enzyme activity occurring in cercariae exposed to red light. These enzymes are involved in the process of signal transmission from photoreceptors in the eye and energy release from electron transport systems, respectively, which ultimately increases cercarial movement in searching for a suitable settlement site.

The kinds of colour preferences demonstrated by these trematodes are in contrast to many species of marine invertebrate larvae that typically settle on darker, deep colours and less-reflective substratum that may provide greater shelter and concealment from predators (Satheesh and Wesley, 2010; Dobretsov et al., 2013). Little is known about the colour preferences of free-living metacercariae that utilize other types of transport hosts, such as molluscs and crustaceans, although *Himasthla rhigedana*, a species that mainly utilizes crabs as transport hosts, has been found to predominantly encyst on black rather than white or grey surfaces (Zimmer et al., 2009). Nevertheless, it seems likely that these kinds of species will demonstrate colour preferences that in general differ from those utilizing plants and probably reflect conditions found in the habitats where such hosts typically occur.

3.3.2.3 Responses to waterbourne chemicals

Dispersing cercariae may also respond to waterborne chemical signals. Krull and Price (1932) found that if a washed pebble was rubbed on the back of a frog transport host *Megalodiscus temperatus* was attracted towards it, temporarily attaching and exploring the inert surface before swimming away, indicating a component of the frog's skin mucus was responsible for inducing a chemotactic response in the parasite. In contrast, the dispersal period of *Parorchis acanthus* cercariae was found by Laurie (1974) not to be affected by exposure to casein hydrolysate, suggesting proteolysis was not a mechanism for generating encystment cues in the same manner as it induces settlement responses for certain marine invertebrate larvae (Zimmer-Faust and Tamburri, 1994). This indicates that the encystment rate of this species is not influenced by chemicals generated during general metabolic processes or to digestion or assimilation of proteins acquired by potential transport hosts through dietary sources.

A more detailed study on chemo-stimulants was undertaken by Tripathi et al. (2014) on *Fasciola gigantica*. These plant-targeting cercariae were found to show some degree of attraction towards carbohydrates (starch) and amino acids (proline and serine) with maximum responses being induced by serine after a 15 min exposure. Nevertheless, prolongation of chemo-attraction for 60 min resulted in significantly higher levels of attraction for all chemicals for these older cercariae. However, the greatest level of cercarial attraction was achieved with a combined stimulation of red light and serine (Tripathi et al., 2014) indicating that a mixture of different signals are responsible for the orientation behaviour of these parasites.

3.3.3 Duration of dispersive phase

The duration of the dispersive period is relatively short, at least under laboratory conditions where typically it may last for a few minutes to a few hours (Krull, 1934; Stunkard, 1966, 1967a; Pike, 1968; Coil, 1984). However, some degree of caution needs to be applied to such studies as they were typically undertaken in shallow glass or plastic dishes that presented an abundance of hard surfaces. The majority of free-living metacercarial species appear to respond positively to such conditions which may have initiated premature encystment responses due to the cercariae being overloaded with thigmotactic settlement cues. Certainly under natural conditions such a rapid and indiscriminate reaction upon emergence does not appear beneficial to the parasite, and it therefore seems unlikely that they would adopt such behaviour when the daily numbers of emerging cercariae for these species is comparatively low (see Section 3.2), and therefore the host selection of each individual cercariae becomes more important compared to species that penetrate target hosts and have higher levels of cercarial production.

There is little quantitative data associated with the duration of the dispersive phase and the rate of encystment of free-living metacercariae. Singh (1957) determined that the freshwater amphistome, *Cercaria lewerti*, had a mean dispersive period of 56 min, ranging between 20 and 130 min before encystment commenced, while Lei and Poulin (2011) showed that the marine *Philophthalmus* sp. had a relatively linear encystment rate over time with all cercariae having encysted by 3 h postemergence.

However, the most detailed analysis of encystment rates have been undertaken on the marine species, *Parorchis acanthus*. The success and rates of encystment varied between individual studies suggesting experimental conditions or parasite strain had a strong influence on results. Stunkard and Shaw (1931) reported that *P. acanthus* cercariae spent 6 h dispersing before encystment began, all metacercarial cysts being formed between 6 and 15 h postemergence, with 57.1% successfully encysting. The remaining cercariae continued to be active at least up to 22 h postemergence before all died without attempting to encyst. In contrast, Rees (1937) found that cercariae swam for only 2 h before beginning to encyst, successful cysts being formed between 2 and 4 h postemergence with a 75% success rate. Morley et al. (2003a) also found a more rapid rate of encystment with almost 75% encystment having occurred by 3 h postemergence and a 95% success rate being achieved by 12 h postemergence.

3.3.3.1 Influence of abiotic factors

Environmental conditions such as salinity and temperature influence the speed and success of *Parorchis acanthus* encystment. Stunkard and Shaw (1931) found that reducing the salinity caused an initial increase in the number of successful encystments to 64.7% (75% seawater) and 78.9% (50% seawater), although the rate of encystment was slower over a more protracted period. However, a lower salinity of 25% seawater resulted in a drastic reduction of successful encystment (27.3%) but undertaken at a faster rate, the majority occurring by 1 h postemergence. Rees (1937) reported that increasing the salinity produced more variable results. At 125% seawater, successful encystments dropped to 41.7% and did not commence until 7 h postemergence. However, higher salinities of 150% and 175% seawater produced success rates comparable with controls but occurring at a faster rate at 150% seawater and a slower rate at 175% seawater. No encystment occurred at 200% seawater. A similar response to lowered levels of salinity was also demonstrated by *Philophthalmus* sp. where encystment success was lower and took place over a more extended period compared to controls. Nevertheless, unencysted cercariae remained alive, and it was suggested that lower salinity interfered with the onset of cyst formation by imposing osmotic stress (Lei and Poulin, 2011).

Temperature also affects encystment. Williams (1969), using a strain of *Parorchis acanthus* from Sierra Leone found that the percentage of encystments that occurred within 6 h of emergence increased from 1−2% at 6 °C to 40−60% at 39 °C with the maximum longevity of swimming unencysted cercariae declining from 600 h at 6 °C to 19 h at 39 °C. In an Irish strain of *P. acanthus* encystment success was greatest at 10 °C and 15 °C, being close to 100%, but significantly declined at 25 °C (Prinz et al., 2011). Similarly, in the freshwater species *Notocotylus attenuatus* at temperature of 4 °C the cercariae die without encysting, however between 10 °C and 25 °C encystment remains largely unaffected occurring rapidly within a few minutes (Harris, 1986).

Water turbulence similarly appears to influence the rate of encystment. Mechanical agitation of the water in which cercariae are swimming has been found to induce an accelerated encystment rate in a range of species (Stunkard and Cable, 1932; West, 1961; Stunkard, 1966, 1967b; Murty, 1966). Young (1995) has suggested marine invertebrate larvae may detect turbulence by detecting shear, and the same may be true for cercariae. Certainly there are many disadvantages to remaining in a highly turbulent environment and the ability to detect violent physical forces would ensure that cercariae could encyst rapidly, likely with only a limited regard for transport host choice, in order to ensure that damage was kept to a minimum and thus maintaining at least some chance of transmission to the definitive host. Physical cues from the settlement surface may also influence rates of encystment. LeSage and Fried (2011) found that *Zygocotyle lunata* cercariae responded to different inert surfaces by encysting quicker on 'Styrofoam' and aluminium foil than on glass or plastic dishes, suggesting that weak or inappropriate physical stimulus from potential transport hosts could prolong the dispersive phase.

3.3.3.2 Ambush behaviour during dispersal

Certain species of Philopthalmidae as an aspect of their dispersal phase perform an initial settlement on a surface but do not begin to examine it as a precursor for encystment. Instead, using a sucker in the terminal part of the tail, they 'stand up' on the tail tip. The attached upright cercariae then periodically bend the body, performing circular searching movements known as 'ambuscade' or ambush behaviour (West, 1961; Howell and Bearup, 1967; Rees, 1971; Diaz et al., 2002), and may also perform this behaviour while hanging down attached to the water surface (Diaz et al., 2002). The target host for such cercariae are typically fast-moving crustaceans and this behaviour response is shared by both free-living and penetrative metacercarial species (Prokofyev, 1994; Galaktionov and Dobrovolskij, 2003).

These cercariae respond to vibrations in the water in an undirected manner if the distance to the disturbance is greater than three to four cercarial body lengths. Vibrations at a closer distance initiate an attack response and the cercariae will stretch out in a directed manner towards the source of the disturbance and attempt to attach. Nevertheless, this behaviour pattern appears only effective in environments where water velocities are low and is abandoned in water flows exceeding 8–10 cm/s (Prokofyev, 1994). If attachment to a crustacean is successful, then the cercariae will detach from the bed surface and wrap its tail around the contact point to maintain and increase its purchase before encysting (West, 1961).

3.3.4 Himasthla rhigendana *as a model species for dispersal*

The most comprehensive study of cercarial dispersal and settlement has been undertaken on *Himasthla rhigendana*, a common species found in California salt marshes, which utilizes crabs and snails as transport hosts. Laboratory studies carried out in static water conditions have determined that cercarial vertical distribution is governed by photo-induced positive geotaxis that was independent of light intensity. Horizontal swimming was infrequent compared to vertical movements where the bottom acted as a reflective boundary, causing cercariae to swim up and down in a predictable manner within a narrow water layer above the bed (Fingerut et al., 2003). Varying salinity had no effect on swim speed or direction, but an increase in temperature resulted in a substantial rise in swim speed. Experiments conducted in a flume showed that fast flows caused turbulent mixing which overwhelmed behaviour effects resulting in cercariae being distributed like passive particles throughout the water column. However, in slow-flow conditions cercarial behaviour prevailed and distribution reflected the results found in the still-water experiment (Fingerut et al., 2003). These studies thus conclusively demonstrate that larval behaviour and not passive transport are the principal mechanism determining dispersal of these cercariae in to the 'host space'.

Further studies on this species were conducted under field conditions that determined that cercarial emergence was light stimulated and that maximum settlement occurred within 60 cm of the source snail host, declining to background densities by 450 cm distance. In faster flows, increased turbulence resulted in higher levels of variability in settlement density from the snail host. The actual dispersal potential of cercariae was 130 times less than passive transport primarily due to their active orientation behaviour resulting in downward larval swimming with cercariae that settled farther away doing so due to rejection of their initial touchdown site. Mean downward swim speed was three times faster than turbulent fluctuations indicating that cercariae could overpower eddies to reach the bed and target transport host thus supporting the conclusions of laboratory studies that larval behaviour is the dominant mechanism responsible for dispersal (Zimmer et al., 2009).

4. SETTLEMENT PHASE

4.1 Settlement and encystment

The majority of marine invertebrate species demonstrate the same basic pattern of settlement behaviour. After the period of distribution and habitat selection have been achieved the process of settlement by pelagic larvae could be divided into three steps: attachment, exploration and fixation (Crisp, 1976). Attachment may not necessarily be permanent, but allows the larvae to test the nature of the substratum. If the full settlement responses are elicited, then the larvae become permanently fixed and proceed to metamorphose. However, if the substratum proves to be unsatisfactory the larvae swim off to locate and test other surfaces and are thus able to make repeated and active attempts to find a suitable place for development and growth (Crisp, 1976).

Free-living metacercariae demonstrate a similar process of settlement responses with the behaviour of cercariae upon approaching a potential settlement site retaining a remarkable consistency across species (Dönges, 1962; Dixon and Mercer, 1967; Howell, 1983). Once swimming cercariae are within a few millimetres of the transport host surface they tend to continue to swim within close proximity for a few seconds before making an initial contact using attachment structures such as suckers or locomotory appendages. If the sensory feedback the cercariae receive is satisfactory, it may begin to crawl over the surface, presumably examining the substrate for a suitable encystment site. At this stage, if insufficient settlement/encystment cues are received by the cercariae it will detach from the surface and swim off to examine another potential encystment site. Temporary attachment remains poorly understood for these species, and it remains unknown if the bioadhesive secreted by some cercariae during temporary attachment is the same as that laid down during permanent fixation and encystment. Marine invertebrate larvae such as barnacles are capable of secreting both a temporary bioadhesive and permanent cement (Robson et al., 2009) and the same may be true for trematodes. Certainly, these parasites have gland cells, secretions and attachment organs that are considered to impart a temporary adhesive function, although the properties of the secretions providing the adhesion have received little attention (Whittington and Cribb, 2001).

If the quantity and quality of the settlement/encystment cues are satisfactory the cercariae will begin the process of encystment. The parasite initially twists and turns on the spot to correctly orientate itself and present its ventral side towards the surface from which bonding cement will be released. Cytogenous secretions covering the entire body are then extruded forming the multilayered protective cyst, and the tail is shed (Figure 1).

4.2 Gregarious and substrate-associated settlement behaviour

The spatial distribution of marine invertebrate larvae after settlement is aggregated on specific sites. The settlement responses that drive this distribution have been extensively studied and the formation of colonies due to chemical cues, in particular, can be separated into three categories, namely those in which larvae respond to the presence of their own species (gregarious settlement), and substrate-associated responses that include those where larvae respond to unrelated species such as a basibiont (associative settlement) or those responding to microbial films covering a substrate (biofilm settlement). These responses have the same two advantageous results for the organism: re-aggregating populations after dispersal thereby improving reproductive efficiency and clustering individuals in suitable habitats that may increase long-term survival (Crisp, 1976; Hadfield and Paul, 2001).

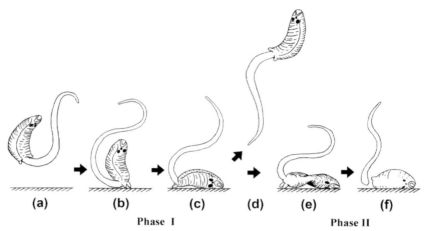

Figure 1 Typical cercarial settlement behaviour as demonstrated by *Notocotylus ralli*. Phase I. Initial contact: (a) approaching transport host surface and swimming across it in close proximity; (b) temporary attachment to substrate using locomotory appendages or other attachment organs; (c) crawling and inspection of transport host surface; (d) detaching and swimming away from unsuitable surface. Phase II. Permanent fixation: (e) after successful identification of a suitable encystment site the cercariae twists and turns on the spot to correctly orientate itself and present its ventral side to the surface; (f) cercariae encysts and the tail is shed. *Redrawn and modified from Dönges (1962) with kind permission from Springer Science and Business Media.*

The structure of free-living metacercarial cyst populations suggests that similar levels of interactions occur to form the clusters of parasites found on transport hosts. Two types of settlement responses for metacercariae appear to exist: a population-level response, where metacercariae will occur in high densities, and an individual-level response, where metacercariae will encyst in close proximity to another. In both cases the accumulations of metacercariae may improve transmission efficiency to the target definitive host by creating high-density populations on sites upon the transport host that are most likely to be ingested or grouping individuals close together which may improve their long-term viability. Unfortunately, no standard terminology has been devised for describing these two levels of interactions and therefore the parasitology literature contains a diverse range of definitions. Therefore, for the purposes of this chapter I use the following most common terms —'cyst aggregations' for population-level responses and 'cyst associations' to describe individual-level interactions.

4.2.1 Cyst aggregations

The aggregation of metacercarial cyst populations at specific localities on the surface of transport hosts is a common feature of many species. Such nonrandom distribution may be due to hydrodynamically sheltered localities, attraction to or repulsion by certain host cues, antifouling excretory/secretory products released by the host or by interactions among the same or different species of free-living metacercariae or other kinds of epibionts.

Aggregated populations have been reported for both animal and plant transport hosts (Krull and Price, 1932; Varma, 1961; Probert, 1965; Wetzel and Shreve, 2003), although they do not exclusively represent the only sites upon the host where metacercariae may encyst. Certainly, it would appear that once the metacercarial population reaches a specific level of density on the chosen locality of the transport host, encystment will 'spill over' into other areas, generally close to the site where the cyst aggregation occurs (Krull and Price, 1932; Wetzel and Shreve, 2003).

From the above studies it seems that the formation of such aggregations is more likely influenced by substrate-associated rather than gregarious cues, certainly in the early stages of colonization where the population density of metacercariae will probably be too low to generate a strong level of response for conspecifics. However, Neal and Poulin (2012) found that *Philophthalmus* sp. cercariae preferred sites on snails with preexisting cysts and that this inclination increased with increasing cyst density and they also did not avoid preferred sites that were already crowded with cysts. Such behaviour would suggest gregarious cues were more important than substrate-associated ones for the aggregation of this species. Thus, the prominence of any one particularly category of settlement cue for inducing aggregation may be species specific.

4.2.2 Cyst associations

Cyst associations may be a relatively common feature of many free-living metacercariae species but remain infrequently described, although certain species such as *Parorchis acanthus* have been found not to associate under laboratory conditions (Morley *unpublished observations*). Certainly from imprecise descriptions in the literature it is not possible to differentiate between aggregations and associations in many cases. Nevertheless, it is apparent that they occur in a range of families, utilizing different transport hosts (Burns, 1961; Graczyk and Shiff, 1994; Abrous et al., 2001) and may also be found in free-floating metacercariae where they may form into long floating strands of up to 5 mm in length containing as many as 500 metacercariae (Holliman, 1961; Stunkard, 1981a), with the size and shape of these strand associations being determined by the diameter of the bivalve siphon from which they emerge (Stunkard, 1981a). Similarly, *Megalodiscus temperatus* cysts are only loosely attached to the tadpole transport host and are easily detached. These detached cysts possess a sticky coating which results, at least under laboratory conditions, in them often adhering together forming free-floating associations (Krull and Price, 1932).

For free-living metacercariae that remain attached, associations are generally determined to occur when cercarial encystment takes place in close proximity to another cyst, typically of three or more cysts to eliminate the chance that two cysts may occur together due to a random process. The distance between cysts where associations may be judged to have formed may vary from one species to the next due to the relative size of the cysts produced and the criteria adopted by individual researchers. Abrous et al. (2001) considered associations occurred when the distance separating cysts was 200 µm or less for *Fasciola hepatica* and *Paramphistomum daubneyi*. In contrast, Graczyk and Shiff (1994) regarded associations in *Notocotylus attenuatus* were only present when the distance between metacercariae was approximately equal to double the width of the external cyst wall (20–25 µm), where the mucoid material of the outer cyst layer was spaced over the area between the cysts, a criteria also adopted by Morley et al. (2002) in their study of the same species. This level of close contact between metacercariae is likely to improve long-term viability by the conservation of moister under semiaquatic conditions. Huffman (1986) found associations of *Sphaeridiotrema globulus* metacercariae could be extensive on the inside shells of molluscs, forming sheets, pyramids and other three-dimensional forms. The outer cyst layers of these metacercariae were joined together by a matrix containing calcium, probably of host origin, which also formed an outer coat surrounding those cysts where no contact with another metacercariae occurred.

The formation of associations, at least for *Fasciola hepatica*, has been found not to be influenced by intensity or duration of lighting, water quality or quantity, or the siting of restrained source snail hosts to particular areas in experimental containers. Nevertheless, two water changes a day resulted in a reduction in the number of associations formed (Abrous et al., 2001). Turbulence induced by stirring the water has produced mixed results. Mechanically stirring either once per minute or per 5 minutes did not influence the number of *F. hepatica* associations formed (Abrous et al., 2001), but gentle continuous stirring with a magnetic stirrer for 4 h to prolong the swimming period of *Notocotylus attenuatus* cercariae caused no associations to be formed even if cercarial density was high (≤ 50 cercariae per cm^3) (Graczyk and Shiff, 1994). However, even under the same experimental conditions two species (*F. hepatica* and *Paramphistomum daubneyi*) can demonstrate differences in the frequency of association occurrence and the number of metacercariae found within each association (Abrous et al., 2001) indicating that the occurrence and prominence of this characteristic may be species specific. It seems likely therefore that their formation is predominantly associated with gregarious cues rather than substrate-associated ones. Pollution may also influence their formation (Morley et al., 2002) and is discussed below.

4.3 Settlement and encystment cues

The final settlement cues initiating the process of encystment of free-living metacercariae remain largely unknown. Nevertheless, a large amount of literature exists on both the host signals which trigger attachment and penetration of cercariae into animal hosts (Haas, 1992, 1994, 2003) and the variables that activate the settlement of many species of marine invertebrate larvae (Crisp, 1974; Rodriguez et al., 1993). It seems likely that free-living metacercariae will react to similar signals, with chemical cues and physical factors or processes being the main instigators of settlement. In a similar manner to penetrative trematode species (Haas, 1992, 1994), cercariae of individual free-living metacercarial species will most probably respond to different combinations of settlement cues specifically tailored to induce encystment on transport hosts most commonly ingested by certain groups of target hosts. Nevertheless, in the absence of appropriate settlement cues free-living metacercariae appears to retain the ability to encyst on 'any surface' in a final default characteristic, presumably to ensure that by progressing to the next developmental stage at least some chance of transmission to the target host is created, however remote those chances may be.

4.3.1 Chemical cues

Settling marine invertebrate larvae respond to glycoproteins, as well as ammonia, L-DOPA, catecholamines, GABA, oligopeptides, polysaccharides and polypeptides associated with the substratum. These chemical inducers are typically associated with conspecific individuals, microbial films and prey species (Rodriguez et al., 1993). Similar analogous examples are likely to determine the distribution of free-living metacercarial populations. For example, a number of chemicals associated with the tegument of target hosts are known to induce penetrative cercariae to attach, explore and enter. They include glycoproteins, cercamides, acylglcerols, carbohydrates, L-arginine, fatty acids and cholesterol with most species responding with great sensitivity and specificity to distinct stimuli (Haas, 1992, 1994, 2003). In particular, echinostome cercariae that penetrate into molluscan target hosts approach by chemokinensis, responding to amino acids, urea and ammonia emitted by the snail (Haas, 2003). It is possible that species of free-living metacercariae that utilize molluscan transport hosts may also use the same chemical cues to induce settlement and encystment.

Chemical inducers from 'conspecific inducers' may also be as potentially important for cercarial settlement as they are for marine invertebrate larvae. It has been found that *Schistosoma mansoni*, infecting mammals, have cercariae that penetrate in a group at the same point. This behaviour has been considered to be associated with either a chemical released by the penetrating cercairae or due to the liberation of host skin components as the parasite forces its way into the tegument which act as a stimulus booster to nearby swimming cercariae that have yet to begin infection (Stirewalt, 1971; Haas, 1992). Furthermore, groups of other cercarial species can also form free-floating aggregates or nets that require coordinated movement between hundreds of individuals to assemble (Wardle, 1988; Beuret and Pearson, 1994) and most certainly involve a degree of chemical communication by 'conspecifics' for construction (Beuret and Pearson, 1994). Thus, it seems likely that diffusible chemicals released during the process of free-living metacercarial encystment may be similarly influential in creating the aggregates and associations of individuals found on transport hosts (see Sections 4.2.1 and 4.2.2).

Microbial biofilms are important in the settlement of many marine invertebrate larvae. Settlement can be induced by films of diatoms and bacteria and may, in the latter case, be generated by the presence of extracellular polysaccharides or glycoproteins attached to the bacterial cell wall or soluble compounds released by these films which may also induce searching behaviour on the substratum. Nevertheless, biofilms may also inhibit settlement, and the exact effect it imparts on larvae may be partly determined by its specific composition that can spatially vary widely (Rodriguez et al., 1993). Larvae are able to extract a great deal of information about a surface from the biofilm coating it, including length of submergence, tidal height and local hydrodynamics (Krug, 2006). However, settlement can be inhibited by young biofilms while induced by mature ones. In contrast, it has also been found that for some species biofilm age will not influence settlement but larval ages will; young larvae being inhibited to settle by films but older ones induced (Krug, 2006). Thus, biofilms can have a complex effect on larval settlement behaviour but their influence on structuring free-living metacercarial populations remains unknown, with only the encystment of *Notocotylus ralli* being shown to be inhibited by the presence of an algal biofilm on the substrate (Dönges, 1962). Nevertheless, extensive studies on the interactions between mammalian host intestinal microflora and helminths (Wescott, 1970; Vieira et al., 1998) would suggest that microbial organisms have a similar complex influence on parasite establishment under these conditions and may therefore be capable of having a comparable relationship in free-living environments.

Some marine larvae are induced to settle by chemical signals from potential prey species. Although 'prey' is a concept not appropriate for the settlement of free-living metacercariae, such chemical signals are analogous of those likely generated by living plant and animal transport hosts and may thus form a similar role. Invertebrate larvae can respond to inducing chemicals such as oligopeptides found on the surface of prey plants or to dissolved substances released into the water by prey animals allowing larvae to respond without needing to establish direct contact with the substratum (Rodriguez et al., 1993). Similarly, potential predators may also produce inducing signals to attract larval prey. In contrast, toxic and nontoxic inhibitors of settlement can also be released by potential prey, and in one sense settlement and subsequent establishment of larvae could be the outcome of rejection responses to negative signals rather than the result of an absence of positive responses to favourable signals (Woodin, 1991). Stimulating attachment signals for cercariae that penetrate target animal hosts have been extensively studied (Haas, 1992, 1994) while both cercarial stimulation and inhibition cues have been reported for some plants to *Schistosoma mansoni* (Warren and Peters, 1968). It seems likely that cercariae of free-living metacercariae will possess similar complex neurochemical signalling mechanisms and will thus demonstrate analogous responses to transport hosts chemicals as invertebrate larvae show towards prey species.

4.3.2 Physical cues

Physical cues may be just as important as chemical ones for inducing settlement and encystment of free-living metacercariae, but they remain more difficult to determine, particularly on a living transport host where chemical signals would be expected to predominate. It is unusual for a marine invertebrate larvae to attach and fix immediately it touches a surface (Crisp, 1974), and the same appears to be true for free-living metacercariae (see Section 4.1). Thus, it is to be assumed that mechanical contact is an important stimulus and suitable larval receptors are necessary to interpret the stimulatory signals.

Responses to water currents in contrast are likely dependent on the strength of the flow, although it is unknown if larvae or cercariae are able to appreciate water movements while being carried freely in a current (Crisp, 1974). Only in slow-flow conditions are cercariae able to overcome water currents and encyst (Fingerut et al., 2003), suggesting rheotaxis is probably not occurring. However, responses to turbulence caused by mechanical agitation will result in a rapid encystment by cercariae (see Section 3.3) indicating that parasite responses to violent forces occurring in the vicinity of a potential transport host may promote a shortening of the usual exploratory phase and an immediate encystment in order to protect the cercariae from physical damage.

Surface textures and contours determine the topography that larvae encounter while approaching a settlement site and which they appear particularly sensitive to. Most marine invertebrate larvae settle more readily on rough rather than smooth surfaces (Crisp, 1974), but the opposite is true for cercariae who prolong their dispersive phase or completely avoid transport hosts with substrate that is excessively textured (Pecheur, 1967; LeSage and Fried, 2011) probably because the ventral side of the cyst is unable to form a firm contact for adhesion.

On a larger scale many larvae can locate and settle in surface irregularities and concavities due to the advantage of shelter which they offer (Crisp, 1974). Such behaviour is termed 'rugotropism' and for large number of species it appears to be mediated by tactile rather than visual responses. Similar kinds of encystment behaviour are apparent for some free-living metacercariae on plant hosts (Varma, 1961; Pecheur, 1967). However, surface projections usually have the opposite effect causing the larvae to swim off or settle a short distance from it. Thus, surfaces with a fibrous texture may discourage settlement (Crisp, 1974).

In general, cercarial tactile responses have rarely been considered for free-living metacercariae. Dönges (1962) considered thigmotaxis (responses to physical contact) was responsible, along with phototaxis and geotaxis, with the encystment behaviour of *Notocotylus ralli*. Dixon and Mercer (1967) regarded that the attachment phase of *Fasciola hepatica* cercarial encystment was initiated by tactile stimulus which produced a firm adhesion by the action of both suckers. They believed tango receptors located near the oral sucker may initiate this step. Nevertheless, species that encyst on plants have fewer sensory receptors than those that encyst on animals (Tolstenkov et al., 2012) suggesting that a greater range of orientation responses are required to settle and initiate encystment on these more active organisms.

4.3.3 Encystment cues for free-floating metacercariae

The cues used by free-floating metacercariae to initiate encystment remain more difficult to determine. Monorchiidae species encyst before being expelled from the bivalve host and the main drivers of this behaviour are likely to be chemical cues associated with this mollusc. Graefe (1971) has demonstrated that the haploporid *Saccocoelioides octavus* responds to changes in the osmotic pressure as the cercariae emerge from the snail host into the freshwater environment, being unable to encyst in a 0.45% saline solution, and this cue may be applicable to other free-floating freshwater haploporid species.

The cues determining the formation of free-floating cysts in those species that demonstrate it as a secondary characteristic, such as *Fasciola hepatica*, are more difficult to determine. Although a small number of factors can influence the proportion of emerging cercariae that encyst in a free-floating manner (see Section 4.6), the main trigger remains unknown. Certainly, it seems unlikely chemical cues may instigate encystment, although responses to excretions from floating microbial organisms cannot be completely discounted. Physical cues upon contact with the water surface or floating particles could induce an encystment response, although it is also possible that no positive cues will initiate this behaviour, rather the inability of cercariae to detect suitable settlement cues from transport hosts within a certain timeframe may result in some cercariae encysting in a free-floating manner.

4.4 Metacercarial cyst morphology

4.4.1 Morphology of fixed metacercariae

The majority of free-living metacercariae reside in a protective cyst which is formed by secretions from the cercarial cytogenous glands. During the process of encystment the cercariae sheds its tail, and cystogenous material is released outwards from the tegument where it polymerizes forming a protective envelope. The cyst shape may be spherical, hemispherical, ovoidal or flask-shaped, with the ventral aspect of the cyst conforming to the curvature of the surface on which encystment takes place (Fried and Grico, 1975; Fried, 1994; Galaktionov and Dobrovolsskij, 2003). However, other modes of cyst attachment can also occur. For example, *Saccocoeliodes pearsoni* anchors itself to filamentous algae by a narrow stalk made up from the cytogenous material, with the body of the cyst hanging down from this stem (Martin, 1973).

Cysts are multilayered with each layer differing in structure and chemistry. The number and composition of layers varies between species which may reflect the environmental conditions (freshwater, marine or semiaquatic) under which they must protect the metacercariae. Rees (1967) considered that the cyst wall of the marine species *Parorchis acanthus* did not need to be as resistant as that formed by the semiaquatic *Fasciola hepatica*, as the risk of desiccation persisted for a much shorter period during low tides and the transport hosts, molluscs and crustaceans, could move to more sheltered positions when necessary. Similarly, the outer tanned coat of the *F. hepatica* cyst wall is considered to protect the metacercariae from fungal and bacterial attack (Dixon, 1965), an event considered less likely to occur in a marine environment (Rees, 1967), but reported to have resulted in the disintegration of the outer layer of the cysts of the freshwater species *N. urbanensis* (Singh and Lewert, 1959). Certain species, such as Notocotylids, have cercariae with bodies darkly pigmented with melanin (Nadakal, 1960), which is retained by the encysted metacercariae (Stunkard, 1960; Pike, 1969; Smith and Hickman, 1983). This pigment is considered to offer some degree of protection from UV rays and desiccation (Nadakal, 1960), and may thus help prolong the long-term viability of the metacercariae in the environment.

In many species a ventral plug is found in the cyst wall which during excystation dissolves, possibly due to the secretions of an enzyme by the metacercariae, forming a hole through which the parasite escapes (Dixon, 1966; Pike and Erasmus, 1967). The plug occurs on the ventral side of the cyst and has a different chemical and structural composition to other parts of the cyst wall, but is not exactly morphologically comparable between different species (Dixon, 1965; Pike and Erasmus, 1967; Rees, 1967). Other species can have a lid rather than a plug which opens out allowing the metacercariae to escape (Fried et al., 1978), or in the case of *Philophthalmus* spp. retain a permanent opening in the cyst after its formation. For these species the cyst forms a flask-shaped structure around the cercarial body and when the tail detaches it leaves an open end at the neck of the flask. This opening ensures that the parasite may excyst quickly in the throat of the bird definitive host before destruction by the harsh conditions of the digestive system and also means that the survival of the metacercariae in the environment is much reduced compared to other species (Nollen and Kanev, 1995). Unusually, *Sphaeridiotrema pseudoglobulus*, a species that encysts on the inner surface of snail shells, has a plug on the dorsal rather than the ventral side of the cyst (Lepitzki and Bunn, 1994). Unlike encystment on the outside of snail shells, these cysts may often be partially encased by shell overgrowth (Diaz, 1980; Lepitzki, 1993) which may interfere with the metacercariae's ability to excyst if the plug was positioned ventrally.

4.4.2 Morphology of floating metacercariae

Species which are capable of producing free-floating metacercariae, particularly where it is expressed as a secondary characteristic, also form distinct morphological features. For example, *Fasciola hepatica* creates a collar that allows a bubble of air to be trapped in an indentation which, along with air-filled lacunae in the outer cyst wall, aid floatation (Sinitisin, 1914; Esclaire et al., 1989). Some Haploporid species have been found to develop long lateral filaments extending out from the cyst wall of their free-floating metacercariae (Shameem and Madhavi, 1991; Diaz et al., 2009; Alda and Martorelli, 2014), which may help them to become entangled in the filamentous algae that form the diet of the fish definitive host (Shameem and Madhavi, 1991). Others, such as *Saccocoelioides octavus*, upon encystment in the water do not shed their tails, which remain both attached to the newly formed cyst wall and active (Graefe, 1971). Although it has not been determined, if all nerve connections between body and tail are broken upon encystment, the likelihood is that this is the case, as found for species encysting on the surface of a transport host such as *F. hepatica* (Dixon and Mercer, 1967). Movement of the tail is therefore certainly independent of the metacercariae and maintained due to endogenous contraction of muscles in the tail facilitated by the large amount of glycogen stored in this organ. This activity will probably aid in attracting the attention of the predatory definitive fish host. In contrast, syncoelid metacercariae do not encyst but have bladder-like floatation devices and byssal threads as an aid to attachment to the gills of the target definitive fish hosts (Gibson and Bray, 1977).

4.5 Transport hosts

4.5.1 Host selection

The selection of a suitable transport host to encyst upon is a key aspect of the formation of free-living metacercarial populations. However, it has surprisingly been only infrequently studied, despite representing the most obvious source of evidence for determining cercarial selective behaviour.

Within marine coastal habitats a majority of the free-living metacercariae utilize seabirds as definitive hosts, and thus cercarial transport host selection should reflect the main diet of these animals. The notocotylid *Paramonostomum chabaudi* is an adult mainly in oyster catchers and mallard ducks (Evans et al., 1997). Under laboratory conditions James (1971) found that most of the cercariae of this species encyst within 3 h, preferentially on empty snail shells and inanimate surfaces (rock and glass) with only a few cysts occurring on seaweed and crabs. After 6 h there is a substantial rise in the number of metacercariae found on seaweed. Under conditions of increased or decreased salinity few cercariae were able to encyst, and those that were did so with less specificity although the majority still occurred on empty snail shells and inanimate surfaces, but with an increased proportion utilizing seaweed. At elevated temperatures of 30 °C even fewer encystments occurred with a more even distribution between shells, seaweed and inanimate surfaces (James, 1971). Nevertheless, the choice of the transport hosts utilized in this experimental study, particularly the use of only empty snail shells which will be unable to present the same range of cercarial settlement cues as living molluscs, suggests it may have only limited application to natural conditions.

Species of *Philophthalmus* in the marine environment also mature in avian hosts and a detailed study of their transport host selection was undertaken by Neal and Poulin (2012). Distribution of cysts on transport hosts sampled during the summer from the New Zealand coast was compared with laboratory conditions. In both cases metacercariae preferentially encysted on the shells of live snails. Under natural conditions only a single crab was utilized as an additional transport host while under experimental conditions a much wider additional host selection occurred with cysts found on bivalves, plants and rocks. Although these results may only reflect the preferential conditions for host selection found in the laboratory, it may also be associated with the summer only sampling undertaken in the field (Neal and Poulin, 2012), and therefore the potential for host choice to fluctuate seasonally cannot be ignored.

Under freshwater conditions, the metacercariae of *Notocotylus attenuatus*, a parasite that sexually matures in waterfowl, were found to preferentially encyst on the snail *Radix peregra* (58%), the majority of the encysting population preferring the shells of those individuals that also harboured the intramolluscan stages of this trematode (35%). The remainder of the cysts occurred mainly on the side of the experimental container (17%), empty snail shells (15%) and floating weeds (7%); with only a small number settling on the prosobranch *Bithynia tentaculata* (4%) and no cysts settling on stones. The predominance of metacercariae on snails and their shells matches the preferred diet of the waterfowl definitive host (Harris, 1986).

Interestingly, *Fasciola gigantica*, a species targeting herbivorous mammals, was found to encyst on the sides of glass experimental containers (47%) in preference to grass (35%) or snail shells (14%). Nevertheless, the density of cysts was higher on grass (18/cm^2) and snail shells (8/cm^2) compared to the container sides (4/cm^2) (Cheruiyot and Wamae, 1990) suggesting that grass may indeed be the main host of choice. Such results may reflect the limited surface availability of grass, the presumed preferential transport host, under these experimental conditions. Once the metacercarial population reached a specific threshold density it is possible that other less favourable transport hosts (i.e. snail shells) were utilized, before cysts began to settle on the sides of the glass container. These results highlight how difficult it can be to undertake and interpret such choice experiments under laboratory conditions.

Crisp (1976), in order to explain the power of choice by barnacle cyprids, suggested a form of instinctive behaviour in which close searching and inspection were considered as initial steps in the pattern leading to the ultimate act of settlement. An encounter with a specific stimulus would lead to settlement only if the threshold for releasing the settlement process had been lowered by a period of exploration with several previous encounters with suitable stimuli. It had been found that the rate of settlement, when tested against a standard stimulus, increased the longer the period that the cyprids were denied access to an appropriate releaser, thus implying that the threshold is lowered with increasing age. Therefore, during the exploratory phase the larvae would be stimulated at first to settle only in the most favourable sites, but as exploration continued and the threshold stimulus fell, less attractive sites would eventually release the settlement response. This process would give the appearance of 'choice' but is in reality an effect resulting from the distribution of settlement thresholds in the natural population combined with a fall in the response thresholds with time. It remains to be determined if a similar situation may be influencing transport host encystment distribution for the cercariae of free-living metacercariae.

4.5.2 Molluscan transport hosts
The shells of molluscs commonly provide a hard substratum for the settlement and establishment of a wide range of organisms, particularly in soft-bottomed aquatic habitats where the availability of hard substratum may

be limited (Guitierrez et al., 2003; Abbott and Bergey, 2007). As molluscs also form the principal diet of a range of predators, particularly aquatic birds, their shells have also been used as transport hosts by a large number of free-living metacercarial species. The traits of each individual shell influence its suitability as a substrate for colonization. Shell size is the most important, with larger shells capable of supporting a higher population density and species richness. Nevertheless, the textual characteristics, shell damage, along with the type and degree of shell ornamentation can influence the extent of colonization (Gutiérrez et al., 2003). The spatial arrangement of molluscs in a habitat can also influence the degree of colonization. Although molluscs can be solitary they are more often aggregated in certain locations increasing the extent of shell substrate available for colonization compared to isolated shells and increasing the likelihood of epibiotic settlement (Gutiérrez et al., 2003).

4.5.2.1 Location of metacercariae on host

Metacercarial encystment may take place anywhere on the surface, both on the outside or the inside of the shell, where cercariae achieve entry by pushing back the edge of the mantle next to the shell, forcing their way between the two without penetrating any tissue (Burns, 1961). Individual species however favour specific locations such as near the shell mantle, apex, valve rim of bivalves or on the operculum of prosobranchs (Harper, 1929; Lepitzki and Bunn, 1994; Prinz et al., 2011), although settlement location choice on the host may vary between naturally and experimentally exposed snails where laboratory conditions can result in a much wider range of host location sites (Neal and Poulin, 2012). Most species demonstrate wide transport host specificity, but in general only a few species are preferentially favoured. It is often the case that the preferred snail species for metacercarial encystment was also utilized as the first intermediate host, although it has been disputed whether individual snails that are actually infected with intramolluscan stages will also harbour a disproportionate lower or higher number of metacercariae on their shells (Harper, 1929; Harris, 1986). Nevertheless, the fact that infections with intramolluscan stages can interfere with specific host defence mechanisms against epibionts, resulting in their shells being especially likely to be fouled by micro-epibionts (Mouritsen and Bay, 2000), although unaffected by the presence of certain macro-epibionts (Thieltges and Buschbaum, 2007), would suggest such individuals may potentially be more likely to be colonized by metacercariae. However, Wetzel and Shreve (2003) found a reduced encystment rate on first intermediate hosts *Elimia livescens* by *Macrovestibulum obtusicaudum* metacercariae compared to individual snails not harbouring intramolluscan stages, while Harper (1929) found that *Notocotylus seineti* infected first intermediate host snails rarely harboured metacercariae, which he considered may have been associated with the infection causing a reduced covering of host mucus on the shell, the presence of which appeared to facilitate encystment of this species. In contrast, Harris (1986) found that *Notocotylus attenuatus* had the highest encystment rates on the shells of *Radix peregra* harbouring their intramolluscan stages, a finding also supported by Radlet (1978) who found a single snail of 13.6 mm length harboured as many as 1652 metacercariae. The empty shells of dead molluscs may also be regularly colonized by free-living metacercariae (Stunkard, 1960; Harris, 1986; Prinz et al., 2011), although colonization rates are much lower than that found for live molluscs.

4.5.2.2 Field studies

Studies on the seasonal dynamics of most kinds of metacercarial occurrence on molluscan transport hosts are limited. Notocotylids on *Hydrobia* spp. have been found to have a variable intensity according to location with 10–25 cysts per snail on the White Sea coast (Stunkard, 1967b) but a mean intensity of only 1.3 per snail in salt marshes of Norfolk, UK (El-Mayass, 1991). Within UK freshwater habitats Notocotylids are most abundant on Lymnaeid species during the summer months, but could still be found during the winter, with a peak prevalence of about 60% occurring in June and July and an intensity ranging from 3 to 20 cysts per snail (Harper, 1929).

In contrast, *Philophthalmus* sp. metacercariae were found on the coast of New Zealand during the summer to preferentially occur on four species of snail. Species of *Diloma* had prevalences in excess of 40% but *Cominella glandiformis* and *Zeacumantus subcarinatus* had lower prevalences of 13.2% and 24.6%, respectively. Intensity of infection was more variable for *Diloma* spp. and *Z. subcarinatus* having mean intensities ranging between 1.53 and 1.76 with maximum numbers of cysts per snail of about 4–5, but *C. glandiformis* retained a higher mean intensity of 3.33 and a maximum number of cysts per snail of 8 (Neal and Poulin, 2012).

Nevertheless, the most detailed ecological studies have been undertaken on *Sphaeridotrema* spp. Metacercariae of *Sphaeridotrema pseudoglobulus* in southern Canada were found on 18 of 21 snail species with *Bithynia tentaculata*, a species that also acts as first intermediate host, being considered the main source of infections to definitive duck hosts at most sites. Infection levels varied within snails between sites and among species within individual sites. In general, seasonal differences in prevalence and intensity of metacercarial infections occurred among age classes of *B. tentaculata* at some sites with the youngest snails having the greatest infection levels from the mid- to late summer, intensity levels increasing into the autumn, suggesting that exposure time was a key variable of infection, ageing snails continuing to acquire additional metacercariae. Other snail species appeared to follow a similar pattern. In most sites relatively few overwintering metacercariae were found suggesting that this stage was not the principal mechanism for continuing infections from one growth season to another. Analysis of the data suggested that the importance of any one snail species in the transmission of *S. pseudoglobulus* metacercariae may vary spatially and temporally (Lepitzki, 1993). This assumption was supported by further work using sentinel snails (*B. tentaculata*) to assess transmission of *S. pseudoglobulus* cercariae (Lepitzki et al., 1994). Furthermore, all size classes of *B. tentaculata* were found to be equally susceptible. Mean abundance of metacercariae was negatively correlated with water depth, turbidity and surface pH, and positively associated with surface and bottom temperature. Differences in metacercarial infections of snails occurred at the microsite scale of meters and are likely to be associated with swimming behaviour of cercariae and the movement of the snail host, both of which are considered to have a restricted distribution.

A similar spatial and temporal variability in metacercarial infections of *Bithynia tentaculata* by the related species *Sphaeridiotrema globulus* occurred in ponds associated with the upper Mississippi river in Wisconsin, USA. Prevalence of metacercariae was always greater than 80%, reaching 100% for most of the summer. Abundance, in contrast, peaked in August/September, about a month after the peak occurrence of the preceding redial stage, before declining into the autumn. However, unlike in Canada, overwintering metacercariae populations were considered to play an important role in maintaining infections in migratory birds in the spring, as abundance did not significantly change between October and May (Herrmann and Sorensen, 2009).

Running water may influence the occurrence of free-living metacercariae on molluscs. Studies on *Macravestibulum obtusicaudum* on the snail *Elimia livescens* in a small Indiana (USA) stream found that infection levels were higher in slow-moving rather than fast-moving sections of the stream. The operculum was the favoured site for encystment for this species, particularly the inner surface, and in fast-flowing sections fewer metacercariae were able to encyst on this site. Despite the difficulties of colonizing hosts in the stream there was no significant increase in the number of metacercariae on individual snails that also harboured an intramolluscan infection of this parasite (Wetzel and Shreve, 2003).

4.5.2.3 Biotic factors influencing encystment success

A number of factors can affect the encystment success of free-living metacercariae on mollusc shells. Encystment of *Parorchis acanthus* on the bivalve *Mytilus edulis* is largely dependent on the filtering activity of the mussels. Cercariae are initially inhaled by the bivalve siphons before they crawl out onto the shell and preferentially encyst along the rim of the valves. When mussel valves are forcibly held shut settlement success significantly declines. Similarly, successful settlement was affected by the presence of barnacles on the shell which filtered and ingested cercariae as well as causing small-scale water turbulence, disturbing the encystment process (Prinz et al., 2011).

The nature of the target transport host population also affects settlement success. Six small mussels attract more metacercariae than one large mussel of the same fresh weight, with an increased recovery rate of cercariae after inhalation from small mussels, possibly because of the increased filtration rates and stronger inhalant current of large mussels (Prinz et al., 2011). Indeed, smaller mussels may in fact be better transport hosts as target definitive bird hosts swallow them whole, but will only selectively feed on the soft tissue of larger bivalves. Furthermore, an increasing density of uniform-sized mussels resulted in more successful settlements but with a decreasing intensity of infection on individual bivalves (Prinz et al., 2011).

In the case of molluscs, the eventual death of the transport host or the encystment of metacercariae on an already empty shell does not indicate the chances of transmission to the definitive host have been extinguished. The majority of trematode species that demonstrate a tendency to encyst on the shells of live or dead molluscs are parasitic in waterfowl. These birds will readily consume both living and dead molluscs (e.g. Hohman (1985)) and infections may thus be acquired through either route. Although this alternative transmission mechanism has been long recognized (Lepitzki and Bunn, 1994) its epidemiological significance has yet to be properly evaluated, even though species will readily encyst on empty shells in the presence or absence of live molluscs (Harris, 1986; Stunkard, 1960; Prinz et al., 2011).

4.5.3 Crustacean transport hosts

Crustaceans are commonly colonized by a range of sessile invertebrates. The presence of epizoites may interrupt streamlining and interfere with swimming, increase the energetic cost of movement, inhibit moulting and potentially raise the risk of predation for crustaceans (Glynn, 1970; Bauer, 1981; Xu and Burns, 1991; Willey et al., 1990). Many decapod crustaceans have specialized structures for grooming the body in order to keep it free of epizoites (Bauer, 1981). Nevertheless, moulting appears to be the most frequent mechanism for removing epibionts (Threlkeld et al., 1993). However, intermoult periods increase with age and size of crustaceans, resulting in higher levels of epibiotic colonization occurring on older animals (Hidalgo et al., 2010), and may also increase seasonally which, in turn, result in seasonal variation in settlement (Eldred, 1962). Some crustaceans appear to be able to adjust their moult cycle in response to elevated risks of microbial infection when exposed to 'pathogen-enriched' conditions (Moret and Moreasu, 2012), and the same may also be possible if the exoskeleton becomes excessively fouled. Where moulting or other defensive mechanisms are unable to prevent colonization, the crustacean may be exposed to increased predation risk due to impaired movement or increased visibility to predators (Willey et al., 1993; Hidalgo et al., 2010).

The role of crustaceans as transport hosts for free-living metacercariae is poorly understood and largely unappreciated. This neglect is unfortunate as there is much evidence particularly from the Philipthalmidae, to suggest many cercariae are morphologically and behaviourally well adapted to settle on amphipod and decapod crustaceans. Certainly species such as *Parorchis acanthus* have almost exclusively been described as encysting on mollusc shells (e.g. Prinz et al. (2011)), with crustaceans only infrequently mentioned as transport hosts (Rees, 1971). However, its tail morphology and cercarial behaviour would suggest it is preadapted to utilize fast-moving transport hosts in preference to more sedentary ones such as molluscs. It would therefore be beneficial to determine if host-choice preference indicated an equal or greater attraction to these more overlooked crustacean hosts for such species.

4.5.3.1 Location of metacercariae on host

Metacercarial cysts have been reported from a range of crustaceans including decapods, amphipods and ostracods (Harper, 1929; Angel, 1954; Alexeyev, 1962; Smith and Hickman, 1983; Huspeni and Lafferty, 2004). Unencysted syncoeliid metacercariae have also been found attached to euphausiids and copepods trawled from the open ocean (Dollfus, 1966; Overstreet, 1970), although such species are more often considered to be predominantly free-floating (see Section 4.6).

Detailed laboratory studies of crustacean encystment remain scarce. Nevertheless, Fried and Grigo (1975) have determined some important principles of the encystment process with *Philophthalmus hegeneri* on *Artemia salina*. This trematode is capable of encysting on both nauplii and adult crustaceans and appears to encyst nonspecifically anywhere on the convex body surface. Nauplii usually harbour only one cyst, but as many as three on a single individual can occur. The size of a cyst is about half that of the nauplius, and consequently they weigh down the larvae influencing its swimming orientation, such that a cyst on the head caused the host to orientate vertically or obliquely downwards. The burden of the parasite is so great on these larvae that they are often inactive, moribund or dead within 24 h. Adults, in contrast, appear to be largely unaffected by any metacercarial burden with cysts located on either the thorax, abdomen or phyllopodium (Fried and Grigo, 1975). On larger crustaceans, such as freshwater crayfish, the circulation of water past the gills and stenites ensures that many cercariae will encyst on these structures (Macy and Bell, 1968).

4.5.3.2 Field studies

Under natural conditions crustacean transport hosts have been rarely studied in detail, although a number of species have been noted to harbour metacercarial cysts on their exoskeletons (e.g. Owen (1987) and Huspeni and Laffety (2004)). The most detailed study has been that of Stevens (1996), undertaken in the Carpinteria salt marsh, California, USA, upon the encystment of *Himasthla rhigedana* on the carapace of the yellow shore crab, *Hemigrapsus oregonensis*. Over both the 12 month sampling period and between different sites prevalence of *H. rhigedana* ranged from 24% to 99% and abundance from 0 to 700 individuals, although seasonal changes had little influence on metacercarial occurrence. Abundance was both positively correlated with crab size and the density of infected first intermediate host snails.

In freshwater environments it has been found that the highest seasonal intensity of infection of *Notocotylus attenuatus* on amphipods in the far eastern USSR occurred in July when crustacean activity and cercarial emergence were at a peak. By early September the amphipods had undergone ecdysis, freeing themselves of the metacercarial burden on the discarded cuticle and infections consequently drastically declined (Alexeyev, 1962). Annual variations in the occurrence of freshwater metacercariae on crustaceans have also been reported. Macy and Bell (1968) found that *Astacatrematula macrocotyla* on the stenite and gills of the crayfish *Astacus towbridgi* had wide variations in the annual intensity of infections on crustaceans between certain years ranging from several hundred cysts on the most heavily infected individuals in one year to being nearly absent by the following one.

4.5.4 Plant transport hosts

Plants, particularly aquatic and semiaquatic species, are commonly colonized by a range of macro- and micro-epibionts. These organisms can have substantial effects on the plant host causing slower growth, increasing susceptibility to drag or shading, and reducing areas for photosynthesis and gas exchange. They may also make the plant less sensitive to light inhibition or the damaging effects of UV radiation at shallow depths, and may attract or repel potential grazers (Cancino et al., 1987; Biermann et al., 1992; Wahl and Hay, 1995; Rohde et al., 2008). In turn, colonization of plants can be both induced and inhibited by chemical metabolites produced at these surfaces (Steinberg and Nys, 2002).

The ephemeral nature of some plant species, demonstrating temporal changes in abundance, can influence the dynamics of epibiont distribution with variations among leaves with different ages and parts of a shoot. Plant availability can also result in seasonal changes in epibiont colonization on those parts of the plant most susceptible to temporal changes, such as flowering structures, but relative stability on more vegetative plant parts (Casola et al., 1987; Borowitzka et al., 1990; Kouchi et al., 2006).

The dynamics of the relationships between epibiont and plant basibiont are equally applicable to free-living metacercariae. However, the role that plants may play as legitimate transport hosts for metacercariae is frequently ignored, with them often referred to only as 'substrate', the word 'host' cautiously avoided (Odening, 1976). The downgrading of plants in this manner cannot be realistically justified, and, as demonstrated above, is certainly not the case in the more general field of site selection by meroplanktonic invertebrate larvae. As plants present a similar range of settlement cues to larvae as animals do (Meadows and Campbell, 1972) they must therefore be considered as equally important transport hosts for metacercariae.

Plants are widely used as transport hosts in marine and freshwater environments as they are actively consumed by both herbivorous and omnivorous definitive hosts. Within marine habitats little is generally known about the ecology of trematodes utilizing plant transport hosts that target herbivorous fish (Al-Jahdali and El-Said Hassanine, 2012). For those targeting omnivorous bird definitive hosts, plants under experimental conditions appear to function in an auxiliary capacity with the majority of metacercariae encysting on invertebrates such as molluscs and crustaceans (James, 1971; Neal and Poulin, 2012), although field studies would suggest plant utilization by these species may in fact be a lot less common (Neal and Poulin, 2012). Within freshwater environments the majority of work has focused on species of veterinary or medical importance such Fasciolidea and Amphistomatidae. Limited information is available on wildlife species which appear to be either generalists, capable of encysting on both plant and invertebrate transport hosts, with the plants again mainly taking an auxiliary role (Harris, 1986; Peoples and Fried, 2008), or exclusively utilizing only plant transport hosts (Pike, 1968; Probert, 1965; Sudarikov and Karmanov, 1980, Simon-Vincente et al., 1985).

4.5.4.1 Location of metacercariae on host
In general, only certain species of plants are exploited as transport hosts by any individual trematode species. For *Fasciola hepatica*, Pecheur (1967) found that of nine plant species investigated only three were preferred, all having smooth surfaces which allowed for good adhesion. Metacercarial colonization of plants is dependent not only on the host location abilities of the cercariae but also the distribution and mobility of the snail intermediate host. Certain plant species favour both the settlement of metacercariae and the presence of the mollusc (Pecheur, 1967; Rondelaud et al., 2011). Growth of the plant host is also important for *F. hepatica*. Metacercariae may be lifted up from the moist conditions of the surface soil through the upward growth of herbage. Blades of grass can rapidly raise the parasite 15—20 cm above the surface, increasing their availability to grazing animals (Taylor, 1964), and also lowering their long-term viability due to the elevated risk of desiccation. The ability of snails to migrate and maintain themselves in habitats containing suitable plants is a key aspect in determining the level of colonization by these metacercariae (Rondelaud et al., 2005). Similarly, under certain circumstances distribution of metacercariae on an individual plant may also be more dependent on snail host than cercariae. For example, in order for *Fasciolopsis buski* metacercariae to colonize the edible tubers of the water chestnut *Eliocharis tuberosa*, which lies in the mud at the bottom of ponds normally beyond the reach of cercariae, they first require the infected snails to migrate down the inside of the decaying stems of the plant until they reach the tuber allowing the

parasite to emerge and encyst (Barlow, 1923). For *Fasciola gigantica*, metacercariae are capable of encysting within a single dew drop on a plant surface, and thus the semiaquatic snail host has the capacity to deposit metacercariae more than a metre from the water's edge during terrestrial migrations (Porter, 1938).

Specific locations for encystment on transport hosts are always preferred. For example, Pecheur (1967) found that *Fasciola hepatica* metacercariae occurred mainly on the upper leaf surface of some species but the lower surface of others. In contrast, Hodasi (1972) found that *F. hepatica* encysted mainly on the lower surface of submerged leaves of the grass, *Dactylis glomerata*. Green parts of plants rather than brown, decaying parts are also preferred by *Fasciola* spp. (Hodasi, 1972; Ueno and Yoshihara, 1974), and the majority of *F. hepatica* metacercariae occur on the leaves rather than the stems of grass (Hodasi, 1972). Similarly, *Fasciolopsis buski* preferentially encysts on the nut or tuber of water plants (Barlow, 1923).

A vertical distribution of metacercariae on plants growing in water also occurs. The majority of *Fasciola* spp. metacercariae appear to encyst just below the water surface, typically less than 2 cm depth (Sinitsin, 1914; Ueno and Yoshihara, 1974; Dreyfuss et al., 2004; Suhardono et al., 2006a). A small number of *Fasciola hepatica* metacercariae have also been reported to encyst up to 2 cm above the water surface (Sinitsin, 1914; Hodasi, 1972; Dreyfuss et al., 2004) indicating that vertical migration by cercariae out of the main body of water can occur. Porter (1938) observed *Fasciola gigantica* cercariae climbing up plant stems and leaves above the water, but only in the presence of moisture, such as provided by morning dew, a prerequisite also necessary to initiate encystment. Conversely, *Paramphistomum daubneyi* metacercariae were mainly found at lower depths of 4—7 cm (Dreyfuss et al., 2004). Nevertheless, when *F. hepatica* and *P. daubneyi* cercariae were simultaneously introduced together their distribution was affected, more *F. hepatica* metacercariae occurring closer to the water surface and more *P. daubneyi* found at lower depths suggesting the encystment of both species was disturbed by the presence of the other (Dreyfuss et al., 2004).

Water plants may also accumulate free-floating metacercariae carried by flowing water from the habitat where they initially encysted. *Fasciola hepatica* metacercariae have been found to mainly wash up on the submerged leaves, roots and stems of *Nasturtium officinale*. However, as the water cress grows metacercarial survival is poor and after 21 days most viable metacercariae were only found on the still submerged roots (Rondelaud et al., 2004).

4.5.4.2 Field studies

Seasonal and annual occurrence of metacercarial infections of plant transport hosts, with the exception of *Fasciola hepatica*, is poorly known. Populations of *F. hepatica* metacercariae are formed in aquatic or semiaquatic conditions, but as water and moisture levels fluctuate, and the plant transport host grows, over time will often ultimately reside under terrestrial conditions. Temperature and relative humidity are the two main variables that influence the seasonal occurrence of metacercariae on plant transport hosts by controlling the number of complete life cycles possible in any given period of time. Climatic conditions vary from one region to another, and this controls the number of metacercarial generations that can occur in a year. Thus, in eastern and central Europe, northern Asia, and Scandinavia only 0.5 to 1 annual generation of *F. hepatica* metacercariae will occur. Two generations per year are possible in most of the British Isles, the west coast of Europe, and the highlands of tropical Asia, while up to four generations can take place in subtropical Australia and the highlands of tropical countries. However, as many as six annual metacercarial generations of *Fasciola gigantica* may be possible in the tropics (Boray, 1969).

In temperate climates *Fasciola hepatica* metacercariae appear on the pasture in two periods from August to October and from May to June reflecting the twice yearly infections of snails (Torgerson and Claxton, 1999). Large numbers of metacercariae encysting on herbage in September remain alive throughout the winter, with as many as 50% still being viable in March. However, mortality of metacercariae encysting in May or June is much more rapid, with viability falling to 50% after only 3 or 4 weeks (Ollerenshaw, 1971). Harsh winters, with prolonged subzero temperatures, may result in large-scale metacercarial mortality and a corresponding fall in their seasonal availability (Hoover et al., 1984). Variations in the density of metacercariae on pasture in temperate latitudes is considered to be largely determined by the influence of climatic factors on the mortality rate of *F. hepatica* eggs, which control the levels of infections in the snail host (Ollerenshaw, 1959; Ross, 1977; Smith, 1981). In habitats where winter temperatures are milder metacercariae have increasing levels of overwintering survival (Luzón-Peña et al., 1994), and may consequently be of greater epidemiological significance alone for maintaining infections in definitive host populations (Torgerson and Claxton, 1999). In more arid regions, such as southern Africa, that experience distinct wet and dry seasons, the seasonal occurrence of metacercariae is strongly influenced by these dramatic shifts in climate. *Fasciola gigantica* metacercariae appear towards the end of the wet season and can be found on plants throughout the dry season, coinciding with a reduction in grazing areas and temporary sources of drinking water and the resulting concentration of all hosts in a small number of permanent water sources (Pfukenyi et al., 2006). Amphistome metacercariae in these regions experience the same kinds of seasonal occurrence although with distinct peaks in abundance on plants apparent during the very early and late periods of the dry season (Pfunkenyi et al., 2005).

Dreyfuss et al. (2005), undertaking a more quantitative survey, found that within French water cress (*Nasturtium officinale*) beds *Fasciola hepatica* infections were low with a mean metacercarial burden of 2.6–6.3 per bed and a wide annual variation in the number of beds infected. The distribution of *F. hepatica* metacercariae did not appear to favour any one plant species with an almost equal number of parasites being recovered on both *N. officinale* and *Apium nodiflorum*. In contrast, within the same habitat *Paramphistomum daubneyi* clearly preferred *N. officinale* as a transport host where a higher burden of metacercariae (5.6–8.4 per bed) occurred. In the lower Yangtse valley of China, Barlow (1923) found that in a pond used to grow the red water caltrop (*Trapa natans*), *Fasciolopsis buski* metacercarial prevalence on the nuts, which lie just below the water surface, was close to 100% with a mean intensity of 17.9 cysts per nut and a maximum intensity of at least 40 metacercariae on a single nut.

4.5.5 Miscellaneous transport hosts

Other kinds of organisms have occasionally been reported as being utilized as transport hosts, although little detailed information is available. Angel (1954) reported *Parorchis acanthus* on a range of animals in addition to the usual described transport hosts which included cirripede, mainly on the appendages, and insect larvae of the Acalyptrate group where the body surface became so thickly encrusted with cysts that the insect, although still alive, could barely move. It is unsurprising that insect larvae are utilized as hosts as their exoskeletons, along with many other kinds of arthropods, show many of the same surface characteristics as crustaceans and may potentially be an overlooked common transport host for these parasites.

Annelids have also occasionally been used as transport hosts. *Amphitrite ornata* is a polychaete of coastal mudflats where it lives in a stable mucus-lined burrow. Kyle and Noblet (1985) found unencysted Gymnophallid metacercariae firmly attached but inactive on the body wall and tentacles of this species. Infections varied seasonally, peaking in late spring and autumn, with no evidence that the parasite ever penetrated into the host or induced any pathological effect. Nevertheless, the presence of encysted metacercariae on annelids appears to be relatively uncommon. Fried et al. (2008) found that *Chaetogaster limnaei*, an ectosymbiotic of freshwater pulmonate snails, would occasionally have cercariae of *Zygocotyle lunata* attach to their body surface but no successful encystments were observed, suggesting that they may be an unsuitable substrate for cysts.

4.5.5.1 Vertebrate transport hosts

Vertebrate transport hosts are rarely described. Fish would appear to be too fast moving to allow successful encystment on their body tegument, with all trematodes described from the skin being penetrative species residing just below the surface. Amphibians remain the only vertebrate on which any detailed studies have been undertaken. Amphistomes are the only free-living metacercarial species that utilize amphibians and appear to adopt different life cycle strategies according to the availability of hosts. Typically cercariae will encyst on dark-pigmented spots on the skin of adult frogs. Periodically the skin is shed and eaten by the frog where the parasite matures in the intestines (Lang, 1892; Krull and Price, 1932; Grabda-Kazubska, 1980; Bolek and Janovy, 2008). Encystment of *Megalodiscus temperatus* preferentially occurs on the pigmented regions of the dorsal side of the legs, particularly the hind legs, around the forelimbs and pigmented spots posterior to the eyes and does not affect the frog host in any way (Krull and Price, 1932). In the absence of adult frogs, encystment of this species may also occur on the skin of tadpoles, but are only loosely attached to these hosts, being easily dislodged to become free-floating, except along the back and base of the tail where a more firm attachment is possible (Krull and Price, 1932). In contrast, *Diplodiscus subclavatus* cercariae do not react to the presence of tadpoles, encysting only on the skin of juvenile or adult frogs (Grabda-Kazubska, 1980). Where no amphibian transport host occurs these parasites will encyst in the water and become free-floating (see Section 4.6). Looss (1892) certainly believed that frogs did not consume their own skin regularly enough for that to be the main route of infection, maintaining that consumption of free-floating cysts with mud overwinter was the likely route of most infections.

Field studies of these amphibian free-living metacercariae remain rare. Bolek and Janovy (2008) could find no cysts of *Megalodiscus temperatus* on the skin of adult frogs sampled from sites in Nebraska, USA. Similarly, Efford and Tsumura (1969) were also unable to detect cysts on adult frogs from a lake in British Columbia, Canada. Nevertheless, they did detect cysts on the salamander, *Taricha granulosa*. Metacercariae were mainly found on the body, tail and legs of the host but were only recorded between late May and early July where mean prevalences were 9.1—15.8% and intensities ranged from one to five cysts per host.

4.5.5.2 Other mechanisms of transport

Metacercarial encystment on pebbles and stones has occasionally been reported from laboratory studies (Murrell, 1965; James, 1971; Diaz Diaz, 1976; Neal and Poulin, 2012). At first glance such objects would appear to have little use as transport hosts, and it has been suggested such encystment choices represent transmission dead ends (Neal and Poulin, 2012). However, many birds and reptiles along with some mammals will swallow small pebbles and stones that become known as gastroliths or gizzard/stomach stones when recovered from the intestinal tract (Wings, 2007). These stones may be deliberately ingested for a specific function, such as crushing, grinding and mixing foodstuffs or stomach cleaning. Alternatively, they may occur due to accidental ingestions of material attached to swallowed prey or material mistaken as prey. Gastroliths are typically considered to be larger than at least a grain of sand (0.063 mm) but rarely exceed a few centimetres in diameter with varying physical characteristics dependent on their function, rock type, retention time and abrasion rate (Wings, 2007). The utilization of such objects may provide a viable route of infection, dependent on the frequency of ingestion, and metacercarial encystment on these stones should therefore not be automatically discounted. Nevertheless, any exploitation of such objects by metacercariae cannot strictly be considered as transport hosts, rather they should be deemed as 'transport media'.

4.6 Free-floating metacercariae

Free-floating metacercariae are designated as those species whose life histories in natural habitats contain a phase that is not associated with a transport host and can include both encysted and unencysted forms. They can be separated into two groups: those that adopt a free-floating existence as a principal characteristic of their life cycles and those that only undertake it as a secondary option presumably in the absence of suitable transport hosts and their accompanying settlement cues.

4.6.1 Principal free-floating metacercariae

Only a handful of species have been determined to have free-floating metacercariae as the principal characteristic of their life cycles the majority of which appear to occur in the marine environment. These include a number of species of monorchiidae that develop in bivalve molluscs with the cercariae encysting without emerging from the host (Holliman, 1961; Stunkard, 1981a,b; Cremonte et al., 2001). Metacercariae enveloped by a gelatinous covering are expelled by the excurrent siphon either singly or in associations of up to 500 metacercariae. Neither light nor temperature appears to influence the expulsion of metacercariae (Stunkard, 1981a) and eventually they sink to the bottom where the sticky gelatinous covering may allow them to adhere lightly to algae or other objects in the water, presumably increasing the chances of them being ingested by the target fish host. Similarly, freshwater haploporid species have either long lateral filaments extending from their cysts which may allow them to become entangled algae that form the diet of the target fish host, or their cercarial tail is retained attached to the cyst wall where its continued independent jerky movement may attract predatory fish (Graefe, 1971; Shameem and Madhavi, 1991).

In contrast, marine syncoeliids also possess a principal free-floating metacercarial stage but unusually it does not encyst. These metacercariae are relatively large, have 'bladder-like' structures as floatation devices and long 'byssal threads' as an aid for attachment to the gills of the definitive fish host (Gibson and Bray, 1977). Development of metacercariae occurs initially within euphausiid crustaceans before their emergence into the water as planktonic organisms. Claugher (1976) has suggested emergence is associated with the vertical migration of the crustacean host, which descend from the ocean surface at dawn, possibly activating the release of metacercariae and preventing them becoming entangled with this source host at the water's surface. Certainly, surface populations of metacercariae must reach locally, and probably seasonally, very high densities as they are commonly found entangled in the legs of sea birds that feed on the euphausiids. These entanglements of metacercariae combine with other debris to form strands of tough fibrous material that create 'trematode anklets' around the legs. Up to 98% of birds may have such anklets that can contain as many as 800 metacercariae each (Claugher, 1976; Imber, 1984; Ryan, 1986). For certain species of seabirds, notably the white-faced storm petrel (*Pelagodroma marina*), these anklets may prove lethal as they can bind the feet together interfering with their ability to move and take-off. Thus, in one incident an estimated 200,000 birds died in a breeding colony of at least a million pairs, suggesting an involvement of around 160 million metacercariae (Claugher, 1976).

4.6.2 Secondary free-floating metacercariae

Free-floating metacercariae are generally encountered as a secondary characteristic. They have been recorded from the Fasciolidae, Paramphistomatidae, Philophthalmidae, Psilostomatidae and Notocotylidae (Looss, 1892; Wisniewski, 1937; Probert, 1965; Velasquez, 1969; Morley et al., 2002; Rondelaud et al., 2004), but may potentially occur in a much wider range of species, as it appears to be a relatively poorly documented phenomenon. Metacercariae can be formed on the surface of the water, around floating particles or grains of silt on the bottom (Velasquez, 1969; Hammond, 1974; Esclaire et al., 1989), and appear to retain similar levels of viability compared to fixed metacercariae. It has been most intensively studied in *Fasciola hepatica* where it was first recorded by Sinitsin (1914). This author described the formation of the cyst on the surface of the water and how if pushed below the surface with a pin a bubble of air will often be trapped in an indentation of the cyst wall formed by a collar. This allows the metacercariae to remain floating, although repeated mechanical agitation will eventually result in the cyst sinking to the bottom. Sinitsin (1914) found that around 6% of emerged cercariae formed floating cysts and suggested infections in some mammals may be acquired solely through drinking water contaminated with such cysts.

This aspect of liver fluke biology was largely neglected until relatively recently. Esclaire et al. (1989) were the first to re-examine this characteristic of the life cycle describing in detail the formation of a flange and air-filled lacunae from the outer cyst wall which enabled the metacercariae to float. Floating cysts were produced at the same time that fixed metacercariae were formed, comprising 6.8% of the total produced, with maximum numbers occurring between midnight and 1 am, and were capable of retaining the ability to float for more than 3 months in stagnant water (Esclaire et al., 1989). Further studies established that photoperiod, light intensity, diameter of experimental container, depth of water, presence/absence of plant media and the number of water changes did not influence the production of floating metacercariae but a diurnal variation in temperature between 12 °C and 25 °C significantly increased cyst numbers compared to a constant temperature (Vareille-Morel and Rondelaud, 1991). Additional work on thermal influences have established that high constant temperatures of 30–32 °C result in an increased proportion of *Fasciola gigantica* free-floating metacercariae compared to those occurring at constant temperatures of 16–18 °C or 23–25 °C (Soliman, 2009). However, cold temperature shock (3 h at 12 °C once a week) did not significantly affect the mean number of free-floating metacercariae produced by either *Fasciola hepatica* or *Paramphistomum daubneyi* (Rondelaud et al., 2013).

Floating cyst production could also substantially vary depending on the species of host mollusc or parasite (*Fasciola hepatica* or *Fasciola gigantica*) and increased in those snails which were of the larger size classes when experimentally infected with miracidia (Vareille-Morel et al., 1994a). In addition, the proportion of cercariae forming free-floating cysts was found to be higher from those laboratory bred colonies of *Galba truncatula* that were derived from natural populations that had a low-frequency encounter rate with *F. hepatica* (Rondelaud, 1993). Studies on the production of cercariae from snails found that the proportion of floating metacercariae was particularly high when emergence was first initiated but declined substantially during successive waves (Vareille-Morel et al., 1994b; da Costa et al., 1994).

Floating metacercariae provide a great opportunity for wide distribution of the parasite beyond the limits of the snail's habitat, particularly in running water. However, both laboratory and field experiments found that more than half of the metacercariae had fallen to the bottom in flowing water with only a small proportion settling on emerged vegetation (*Nasturtium officinale*), most of these dying as the plant grew. Nevertheless, the type of flowing water habitat, ditch or furrow, influenced the distribution of metacercariae. Such results suggest that floating metacercariae may have a more important epidemiological role in stagnant rather than flowing water environments (Vareille-Morel et al., 1993b; Rondelaud et al., 2004).

These above stimuli for forming free-floating metacercariae, however, may not be applicable in all circumstances. Species of *Megalodiscus* appear to only form floating metacercariae in the absence of the adult frog transport host and the presumably chemical and visual settlement cues that it presents (Looss, 1892; Krull and Price, 1932). Similarly, *Parafasciolopsis fasciolaemorpha*, a species infecting elk closely related to *Fasciola hepatica*, will encyst on the water surface in the absence of vegetation. Nevertheless, even when vegetation was present close to half of emerging cercariae will still encyst on the surface of water suggesting that this species is not preferentially stimulated to settle on the underside of vegetation by any characteristic of the leaf surface (Wisnienski, 1937; Sudarikov and Karmanov 1980).

4.6.2.1 Additional mechanisms for forming free-floating metacercariae

Metacercariae fixed to the surface of transport hosts may also become free-floating due to a number of factors. Of these, the accidental ingestion of metacercariae by snails grazing on vegetation is the most extensively studied.

Species of *Fasciola* have been found in the faeces of Lymnaeid snails that fed on vegetation to which they were attached, and despite most metacercariae losing their rough gelatinous outer cyst layer during passage through the molluscs intestines, many retained their viability and developed normally when experimentally infected in mammalian hosts (Taylor and Parfitt, 1957; Rajasekariah and Howell, 1978; Yadav and Gupta, 1988; Yoshihara and Ueno, 2004). Under natural conditions this behaviour could result in redistribution of many of these cysts on to the muddy bottom of aquatic habitats as snail faecal material settles after defecation, supplementing those naturally created free-floating cysts that also accumulate there. Such accidental ingestion by snails may also remove metacercariae that have encysted on mollusc shells. Snails will often graze on the shells of other molluscs feeding on algae or supplementing their calcium diet by ingesting grazed shell fragments (Abbott and Bergey, 2007). This kind of redistribution mechanism by grazing snails has been reported in the marine environment for *Parorchis acanthus* metacercariae, which can occur on bivalve shells in particular. Under laboratory conditions grazing *Littorina littorea* and *Patella vulgata* removed between 78% and 94% of metacercariae encysted on the bottom of plastic dishes with viable parasites being recovered from their faeces (Prinz et al., 2009). However, grazing by *Gibbula umbilicalis* removed only 9% of metacercariae with no intact parasites recovered from the faeces. This may be due to the comparatively small size of this snail species making it incapable of ingesting metacercariae whole (Prinz et al., 2009), and suggest the existence of some important general controlling mechanisms associated with the suitability of a grazing snail to redistribute metacercariae. Nevertheless, the long-term viability of such mechanically created free-floating cysts remains unknown. In contrast, ingestion and passage of metacercariae through other animals may not be as successful. *Fasciola hepatica* metacercariae were found to have a 99% reduction in viability after passing through the digestive tract of Rouen ducks (Samson and Wilson, 1974), suggesting that nonmammalian vertebrates may not be suitable for redistributing metacercariae into other habitats.

Changes in the viability of transport hosts may also result in fixed cysts becoming free-floating. Lutz (1892) found that cysts of *Fasciola hepatica* attached to plant parts which decompose over time results in the metacercariae detaching and sinking to the bottom. Similar species that use amphipods, such as *Gammarus* spp., as a transport host may also at some point become free-floating when the host moults the old exoskeleton containing the fixed cysts, and this decomposes, freeing the metacercariae. However, the carapace of decapod crustaceans, such as crabs, is likely to decompose at a much slower rate suggesting that metacercariae attached to these discarded exoskeletons are less certain to become free-floating.

Under some conditions fixed cysts may not form a firm attachment to the chosen substratum. Metacercariae of *Megalodiscus temperatus* are so loosely attached to tadpoles that they usually drop off or are brushed off soon after they are formed (Krull and Price, 1932). Similarly, Lutz (1892) found that the adhesion of *Fasciola hepatica* cysts to certain substrates is loosened over time, particularly if it is a smooth surface, such that water currents are sufficient to detach the metacercariae. Thus, the proportion of free-floating cysts in a metacercarial population found within any given habitat may vary independently of the availability of fresh recruits formed by actively emerging cercariae.

4.6.2.2 Epidemiological significance

One of the most puzzling aspects of free-floating metacercariae, particularly when it manifests as a secondary life history characteristic, is determining their epidemiological significance. Certainly, it may be tempting to consider such metacercariae will pose only a small risk of infection in target hosts. Indeed, transmission success may potentially be lower than found for metacercariae associated with a transport host; however, it would be presumptuous to assume that free-floating metacercariae will only rarely be the basis of infections, particularly when determining sources in any specific case is largely impossible. The main suggested mechanism of transmission for these floating *Fasciola* metacercariae has been ingestion while drinking from parasite contaminated habitats. In this scenario it is generally considered that metacercariae which remain floating on the water surface will be predominantly consumed (Vareille-Morel et al., 1993b, Rondelaud et al., 2004), as target hosts usually only take water from the top few centimetres, with metacercariae that sink to the bottom becoming beyond the reach of ingestion. However, Lutz (1892) pointed out that metacercarial cysts' in the sediment of shallow water bodies can easily rise again due to their low specific weight. Under the action of drinking, the tongues of animals lap at the water stirring up the bottom sediment, resulting in large quantities of mud being accidentally consumed. Thus, under these conditions even sunken metacercariae can be transmitted to target hosts (Lutz, 1892). Furthermore, where water levels have receded, sunken metacercariae resting on the wet mud will be exposed and may be consumed by ruminants due to their widespread practice of voluntary (geophagia) or involuntary ingesting soil while grazing

(Abrahams, 2005). Indeed, soil may be a better long-term survival medium for metacercariae as it retains higher levels of moisture than found on the surface of vegetation. In addition to ruminants, geophagia has been reported for a wide range of mammals, birds and reptiles (Abrahams, 2005) and this behaviour may result in the acquisition of a number of species of free-floating metacercariae. In a similar manner, the free-floating metacercariae of *Diplodiscus subclavatus* that sink into the bottom mud of aquatic habitats can be accidently ingested by hibernating frog target hosts which feed on sediment as they overwinter (Looss, 1892). Ultimately, the proportion of metacercariae that may be transmitted in such free-floating ways is impossible to determine. Nevertheless, some human trematode infections, after patient interviews eliminated traditional transmission routes, have been assumed to have occurred in no other way (Weng et al., 1989; Esteban et al., 2002; Mas-Coma, 2004), and it therefore seems likely that this mode of infection may be more common than anticipated.

5. ABERRANT FREE-LIVING EXISTENCE

Under certain circumstances metacercariae that normally penetrate and establish in the tissues of a target host may become free-living. This phenomenon has been most widely documented for metacercariae that encyst within a host that dies, where the corpse gradually decomposes and the metacercarial cysts are liberated passively into the water. It has been described for *Paragonimus westermani* from crayfish (Loh et al., 1969), *Clonorchis sinensis* from fish (Komiya and Suzuki, 1964a,b) and *Echinostoma caproni* from snails (Christensen et al., 1980), but is likely to occur in many aquatic host-metacercariae systems. Such a mechanism of metacercarial release is not unexpected under such circumstances; however, further work has established, despite the cyst wall not being designed to protect from such environments, that the parasite can retain viability in the short-term. Survival of metacercariae can range from a few days to a few weeks and are capable under experimental conditions of establishing infections in target hosts (Komiya and Suzuki, 1964b; Christensen et al., 1980). Death of metacercariae, at least in the case of *P. westermani*, has been attributed to the low osmotic pressure of water compared to the tissue fluids where the parasite became encysted (Yokogawa, 1964), and this reason is likely to be applicable for most other species. Although transmission of such aberrant free-living metacercariae has been experimentally demonstrated, it has been questioned whether such infections through drinking contaminated water can be acquired in this manner under natural conditions (Komiya and Suzuki, 1964b). Unlike normal free-floating metacercariae these aberrant forms appear to have more limited capacity for floating (Loh et al., 1969) reducing their abilities to stay near the water surface and lowering their chances of being ingested by target hosts in the short-term period they remain viable. Nevertheless, certain human infections have demonstrated no obvious source of infections from these trematode species (Roque et al., 1953; Nagano, 1964), and therefore suggest drinking water should be considered as more than just a theoretically possible transmission route for these species.

At least one example of active liberation by metacercariae from a dying host has been documented. Unencysted forms of *Clinostomum complanatum* progenetic metacercariae exited *en mass* from fish in a tank after an accidental mortality event. The parasites emerged through the gills of the dying fish and were active and motile on the water bottom for an hour after the event (Rizvi et al., 2012). Although these authors considered the metacercariae left the host in order to locate another one, it seems likely that without the limited protection afforded by a cyst wall the lifespan of these metacercariae, physiologically adapted to existence within the host tissues, is unlikely to stretch further than a few hours. Certainly, without the physical adaptations to actively locate, attach and penetrate a suitable target host as shown by cercariae and miracidia it appears dubious that such metacercariae could successfully continue their life cycles, assuming that such an event occurred under natural circumstances.

A further mode of aberrant free-living existence has been demonstrated by a number of species of echinostomes. These trematodes normally encyst within the tissues of target intermediate hosts, predominantly molluscs. However, encystment on the mucus trails of molluscs in the open water can occur under laboratory conditions (Laurie, 1974; Fried and Bennett, 1979; Christensen et al., 1980), although some species remain incapable of forming such cysts (Fried et al., 1997). Varying levels of encystment success exist between the mucus of different snail species with fluctuating numbers of abnormal, and presumably unviable, cyst formation occurring (Fried and Bennett, 1979). When normal cysts are formed in mucus, however, they retain the ability to infect definitive hosts (Fried and Weaver, 1969; Christensen et al., 1980). It remains to be determined if such encystment occurs under natural condition, and is only likely to be a viable mode of transmission mainly where mucus trails containing cysts occur on vegetation or animals that form the diet of target hosts, presuming that these cysts are retained within the mucus after formation. Occasional instances of echinostome encystment in pond water alone have been recorded (Fried and Bennett, 1979) but not every species are apparently capable of achieving this (e.g. Laurie (1974)), and all the cysts formed appear abnormal (Fried and Bennett, 1979) suggesting they may not be viable.

6. METACERCARIAL BIOLOGY

6.1 Metacercarial viability

Free-living metacercariae do not normally feed and are dependent on their glycogen reserves for survival. Exact levels of glycogen vary from one cercariae to another (Ginetsinskaya, 1988), and the same is likely to apply to metacercariae. Such differences arise during development in the molluscan host (Ginetsinskaya, 1988) and when combined with differing intraspecific and interspecific metabolic demands caused by swimming activity and length of dispersal prior to encystment, may produce wide variations in the survival capabilities of individual metacercariae. The process of encystment is, in turn, an energy-dependent activity (Vernberg and Vernberg, 1971; Erasmus, 1972) which will further deplete glycogen levels and influence survival potential. The parasite remains metabolically active after encystment but at a substantially lower level than that found for miracidia and cercariae. The intensity of most metabolic reactions decreases with time in metacercariae due to the gradual reduction in glycogen content (Humiczewska, 2004), and correspondingly the viability of metacercariae declines with age (Ollerenshaw, 1971; Griffiths and Christensen, 1972; Asanji and Williams, 1985), although separate strains of the same parasite species may possess differing levels of metacercarial viability over time (Nollen et al., 1985). Changes in abiotic conditions of the environment may also influence metabolic rates and thus the duration of the survival period (Humiczewska, 2004).

6.1.1 Abiotic effects: temperature

The most important abiotic factor affecting metacercarial viability is temperature. Freezing below 0 °C will often render metacercariae completely unviable (Boray and Enigk, 1964; LeSage and Fried, 2011), although it has been suggested that such temperatures do not kill the parasite but induce irreversible morpho-physiological changes resulting in their inability to successfully infect target hosts (Boray and Enigk, 1964). Nevertheless, these effects appear to be dependent on the kind of subzero temperature exposure. Sudden, gradual or intermittent exposure of *Fasciola hepatica* metacercariae on damp filter paper to −20 °C rendered them unviable. At −10 °C metacercariae were able to retain infectivity for up to 28 days but was more rapidly reduced if the parasites were frozen in water, few remaining infective beyond 7 days. Fluctuating daily temperatures between −5 °C and 10 °C resulted in a high proportion remaining infective beyond 70 days exposure. In contrast, *Fasciola gigantica* metacercariae were more sensitive to subzero temperatures, being all unviable after 30 days exposure to −2 °C (Boray and Enigk, 1964).

Metacercariae maintained in cold temperatures (4−7 °C) generally survived for longer periods than those kept at more typical summer temperatures (20−25 °C). Infectivity was retained by at least some metacercarial species for over a year at cold temperatures but often declined to zero after a few weeks at 20−25 °C (Boray, 1963; Griffiths and Christensen, 1972, 1974; Nollen and Kanev, 1995; LeSage and Fried, 2011). Similarly, species of philophthalmid metacercariae from higher latitudes appear to show a greater level of adaptation to colder temperatures than those from lower latitudes, which may be associated with conditions encountered in their respective natural habitats (Nollen and Kanev, 1995).

In contrast, high temperatures (30−40 °C) substantially reduced the viability period of metacercariae (Boray and Enigk, 1964; Asanji and Williams, 1985; Suhardono et al., 2006b; LeSage and Fried, 2011). However, individual species show variable thermal tolerance to these conditions. For example, *Fasciola gigantica* retained viability for longer at 35 °C than *Fasciola hepatica* which may reflect its geographical distribution in subtropical and tropical latitudes (Boray and Enigk, 1964).

Thermal conditions during development in the molluscan host can also influence the viability of metacercariae. Davtyan (1956) found that rabbits infected with *Fasciola hepatica* metacercariae that had developed in molluscan hosts at 22−23 °C were more viable, measured via pathogenic effects, than those that had developed at 15−17 °C, although parasite intensities remained comparable. However, Chowaniec and Markiewicz (1970) found the opposite, with metacercarial viability being greater from those derived from molluscs at 16−18 °C than at 25−26 °C or 10−13 °C. In contrast, infections in sheep caused only a chronic infection with metacercariae that had developed at 29−32 °C but were lethal from those developed at 23−24 °C (Davtyan, 1956). These studies clearly demonstrate the complex influences temperature may have on the transmission and establishment of metacercariae, and suggest that only limited generalizations can be applied to this abiotic factor.

6.1.2 Abiotic effects: other factors

For many free-living metacercarial species desiccation is another major risk factor capable of reducing long-term viability. It is pertinent particularly for species that utilize terrestrial or semiaquatic plant transport hosts that may be exposed to prolonged dry conditions during times of low rainfall (Thommen and Westlake, 1981; Volaire and Lelievre, 2001), the shells of freshwater pulmonate molluscs which periodically emerge out of water (Green et al., 1992), and invertebrate transport hosts in the intertidal marine environment where long daily phases being exposed to the air may be encountered (Warner, 1977; McMahon, 1990). A reduction in the relative humidity (RH) of the atmosphere to 75–80% at 20 °C resulted in *Fasciola hepatica* metacercariae becoming completely unviable by 10 days exposure, but the same conditions maintained at 10 °C saw some viability retained for up to 31 days exposure. In contrast, *Fasciola gigantica* rapidly lost viability in relative humidity of 90% at 20 °C. The increased susceptibility of this species to desiccation may be associated with its molluscan intermediate host being an aquatic snail rather than, in the case of *F. hepatica*, a semiaquatic snail (Boray and Enigk, 1964).

More detailed studies on *Fasciola gigantica* found that a combination of low humidity (25% RH) and high temperatures (35 °C) resulted in a loss of viability within a week while high humidity (95% RH) and low temperature (20 °C) found viability of exposed metacercariae remaining similar to those immersed in water for over 3 weeks (Suhardono et al., 2006b). Similar reduced levels of viability under desiccation stress were also found for the marine species, *Parorchis acanthus*, although exposure in the shade substantially prolonged viability compared to exposure under direct sunlight (Asanji and Williams, 1985).

In marine environments changes in salinity may often be of equal importance as desiccation. Survival of *Philophthalmus* sp. metacercariae was reduced over time at lower salinities. This suggests that the cyst wall is unable to protect the parasite from long-term osmotic stress either due to an intrinsic inability of the cyst structure to provide adequate protection or that the cyst is not formed properly at low salinities (Lei and Poulin, 2011). In contrast, freshwater metacercarial species are rapidly killed by increasing salinity (Komiya, 1964; LeSage and Fried, 2011). Nevertheless, the viability of metacercariae in the external environment may not always be a sole accurate indicator of their infectivity potential (Prasad et al., 1999) as this is a procedure that is also dependent on a range of biotic factors associated with the host—parasite relationship.

6.2 Metacercarial infectivity

Most free-living metacercarial species are capable of infecting a wide range of similar definitive host species. For example, *Fasciola hepatica* has been recovered from many mammalian species that include ungulates such as wild and domestic ruminants, marsupials, rodents, humans and dogs (Pantelouris, 1965). Others, such as *Zygocotyle lunata* and *Philophthalmus gralli* are capable of excysting and developing in both mammalian and avian definitive hosts, although the success rate is influenced by host susceptibility, strain and age (Nollen and Kanev, 1995; Fried et al., 2009).

However, the most extreme definitive host ranges are found with ectothermic definitive hosts. Amphistome metacercariae that normally infect amphibians have been reported from the intestinal tract of European and Indian pulmonate snails (Honer, 1961; Murty, 1973; Ray et al., in press). Snails are considered to acquire infections while grazing on either vegetation containing fixed metacercariae or on the muddy bottom of habitats, accidentally ingesting free-floating metacercariae (Murty, 1973; Sey, 1983). However, after experimentally infecting snails with *Diplodiscus subclavatus*, Sey (1983) found that the mature parasites produced unviable eggs, indicating the adults were sterile and the snails were simply accidental hosts representing a developmental dead-end. Sey (1983) consequently concluded that all descriptions of adult amphistomes from snail hosts were of a similar nature, an assertion that has yet to be refuted.

Levels of metacercarial infectivity success in laboratory exposures are modest, typically ranging from 7.2% to 75.0% (Macy et al., 1968; Asanji and Williams, 1974; Fried et al., 1978; Prasad et al., 1999), but are likely to be higher than those found under natural conditions. Infectivity can be influenced by a range of factors and may vary according to the age of metacercariae and the original definitive host species (Valero and Mas-Coma, 2000) or even the individual snail hosts used as a source of parasites (Shostak et al., 1993).

In particular the environmental temperature which the metacercariae are exposed to prior to infection may substantially influence infectivity. It is apparent that metacercariae retain the highest levels of infectivity at low temperatures, significantly below the optimum thermal ranges of cercariae and miracidia (Morley and Lewis, 2015). Infectivity then declines as the temperature is raised (Boray, 1963; Chadhri and Gupta, 1985; Ferrell et al., 2001). Ferrell et al. (2001) suggested that the environmental microhabitat choice of cercariae could influence levels of infectivity, with those encysting in locations that experienced cooler temperatures such as shaded locations or greater water depth may remain infective for longer and that natural selection may favour those individuals that encyst on transport hosts with physical characteristics correlated with lower temperatures. Morley and Lewis (2015), taking a broader perspective of metacercarial infectivity considered that this stage was best adapted for maintaining parasites in habitats during colder periods, such as overwinter, when temperatures were too low to facilitate cercarial or miracidial activity. Indeed, under certain natural conditions free-living metacercariae of *Fasciola hepatica* may be the only parasite stage that can retain their viability overwinter in any significant manner (Luzón-Peña et al., 1994).

6.3 Metacercarial excystment

Upon ingestion by the definitive host the encysted metacercariae must become active and escape from the cyst before it can utilize the host as a nutrient source and develop into an adult fluke. Excystment, in a similar manner to encystment, is an energy-dependent activity and this final drain on the metacercariae's remaining glycogen reserves may influence establishment success in the definitive host during this period when the parasite remains nutrient independent.

Two interconnected elements, activation and excystment, are involved in this process. Activation is a behaviour response that is stimulated by both extrinsic (digestive enzymes, pH, temperature, etc.) and intrinsic (parasite secreted products) factors, while excystment is the term applied to the breaching of the cyst and the escape of the parasite into the digestive tract of the host (Irwin, 1997).

The first sensory stimulus the metacercariae receives on being eaten is probably physical shock. During ingestion of the transport hosts as food the parasite is either subjected to mechanical fragmentation before being eaten or undergoes mastication in the mouth. Fish, amphibians and birds may, in addition, swallow food whole, although birds will subsequently ground-up food in the gizzard. This process appears to stimulate many species of metacercariae, which begin to move around inside the cyst and may release secretions related to the disaggregation of the inner cyst wall, as well as also aiding the physical removal of the outer layers of the cyst wall (Irwin, 1997). Following these mechanical influences, if the metacercariae have been ingested by an endothermic host, they are next subjected to a sharp rise in temperature. The role that host temperature may play in the process of excystation is not fully understood for the majority of species (Fried, 1994). For some species, such as *Philophthalmus gralli* or *Philophthalmus hegeneri*, it is clearly the main extrinsic factor (Cheng and Thakur, 1967; Fried, 1981) and is intended to trigger activation and excystment in the host's throat, allowing the metacercariae to escape their unique open-ended cysts before they are destroyed in the stomach, and migrate via the lacrinal duct to the eye orbit (Nollen and Kanev, 1995). For other species, endothermic host temperature ranges of 37–42 °C produce the optimal requirements for excystment under the more direct influence of other extrinsic factors. As the cyst passes into the stomach and duodenum it is exposed to a range of extrinsic conditions, such as low pH and pepsin, which may aid the final separation of cyst from substrate, followed by trypsin and bile salt, which may further stimulate the activity of the metacercariae and digest components or layers of the cyst wall (Sukhdeo and Mettrick, 1986; Irwin, 1997).

The activated metacercariae rotate within the cyst, pushing against the wall and thereby identifying the escape aperture. Many metacercarial species release secretory material at this stage which may be involved in the excystment process by softening the ventral plug region allowing the parasite to escape (Sukhdeo and Mettrick, 1986). Newly excysted metacercariae still retain some degree of the endogenous glycogen stores which they may consume until an exogenous source of nutrients can be utilized (Tielens et al., 1982).

The duration of the excystment process therefore appears to be dependent on a number of factors, associated with fluctuations in the intensity of metacercarial responses to these stimuli. Nevertheless, the most important additional factor is the rate of passage of the ingested contents through the host's alimentary canal, which may vary among definitive host species, and is dependent on the kinds of transport hosts being digested (Asanji and Williams, 1974). Ultimately, a small number of cysts may pass out of the alimentary canal within faeces unencysted. Such cysts may have been trapped in indigestible material and not properly exposed to the necessary mechanical and physiochemical requirements for excystment. Certainly, 100% excystment does not occur under any in vivo conditions (Asanji and Williams, 1974).

7. POLLUTION AND FREE-LIVING METACERCARIAE

Pollution of aquatic habitats is a ubiquitous problem that may have serious implications for the successful transmission of trematodes (Morley et al., 2003b). Free-living metacercariae may be particularly vulnerable to pollutants because both the establishment of populations during the presettlement phase is dependent on correct cues from the transport host surface which may be compromised during pollution events and the protracted periods of weeks and months spent in the polluted environments once encysted on a transport host.

The emergence of cercariae is the first parameter that has been demonstrated to be influenced by pollutant exposure. Survival of molluscan first intermediate hosts can be increased or decreased in freshwater polluted conditions compared to unexposed individuals (Evans, 1982; Morley et al., 2003c; Soliman, 2009), although marine molluscs demonstrate greater levels of tolerance (Morley et al., 2003b). Any change in the viability of the molluscan source host will influence the levels of available cercariae for settlement in the polluted habitat. Cercarial output may often be reduced (e.g. Evans (1982)), with antifouling biocides such as Irgarol capable of completely inhibiting emergence (Morley et al., 2003a), although this negative effect is not universal (Morley et al., 2003a; Soliman, 2009) suggesting species-specific or pollutant-specific effects strongly influence the outcome.

The length of the cercarial dispersive period is changed in polluted conditions with species such as *Parorchis acanthus* demonstrating a much slower rate of encystment and a reduced cyst formation after exposure to antifouling biocides (Morley et al., 2003a). In contrast, vital dye's that are used as antimicrobial pesticides in aquaculture (Plakas et al., 1999) can induce a more rapid encystment response in both marine and freshwater species (Stunkard and Cable, 1932; Stunkard, 1967b; Diaz Diaz, 1976). The proportion of successful cercarial encystments under pollution is also affected. Encystment success is generally reduced, but often only at very high concentrations of heavy metals and antifouling biocides (Evans, 1982; Morley et al., 2001, 2002, 2003a; Bennett et al., 2003; Soliman, 2009). Heavy metal exposure can also cause an increase in the number of free-floating encystments (Morley et al., 2002; Soliman, 2009) but a reduction in the number of cyst associations formed (Morley et al., 2002).

Changes in the viability of metacercariae exposed to pollution only after the cyst has been fully formed have also been recorded. In vitro excystment of *Parorchis acanthus* is generally only negatively affected by the highest concentration of antifouling biocides or heavy metals (Morley et al., 2001; Bennett et al., 2003), although the rates of excystment can be either positively or negatively influenced by metal exposure (Morley et al., 2001). In contrast, the successful in vitro metacercarial excystment of *P. acanthus* exposed to antifouling biocides during the cercarial dispersal phase was much more significantly reduced, indicating a greater vulnerability of the parasite during this relatively short period of a few hours compared to a many weeks exposure of fully encysted metacercariae (Bennett et al., 2003).

In vivo infectivity studies after toxicant exposure of metacercarial cysts have produced variable results. Heavy metal toxicity on *Notocotylus attenuatus* metacercariae had no effect on the number of recovered adult worms from chicks (Evans, 1982). However, exposure to silage waste, fertilizers, or copper resulted in a complete inhibition or reduced worm establishment of *Fasciola gigantica* or *Fasciola hepatica* (Brglez and Wikerhauser, 1968; Gupta and Karma, 1987).

These studies indicate a variable effect of pollutants on free-living metacercarial functional biology. However, an important weakness of these laboratory experiments is that exposures were all undertaken in small plastic or glass dishes, and the unknown influence such conditions may have on parasite behaviour. It therefore remains to be determined as to their relevance under natural polluted conditions. Certainly one important key aspect that is missing from these studies is the manner in which the transport host surface will respond to pollutant exposure, and how this may influence successful settlement of free-living metacercariae.

The tegumental surface and mineral outer shell are the most exposed features of organisms in polluted conditions and the associated toxicant exposure can ultimately influence their structural integrity. Malformations of mollusc shells in polluted habitats have been found to include thickening and changed shape, a decrease in shell hardness, globular malformations on inner shell surfaces, shell chambering and high levels of accumulated heavy metals (Cunningham, 1976; Almeida et al., 1998; Kádár and Costa, 2006; Nuñez et al., 2012). Crustaceans can also bioaccumulate pollutants in their exoskeleton and the regeneration and moulting of this structure can be affected by toxicants resulting in deformed limbs, delayed ecdysis and compromised structural integrity due to reduced levels of chitin (Weis et al., 1992; Gagne et al., 2005; Lewtas et al., 2014). Similarly, while accumulating high levels of pollutants plants may suffer from inhibited growth, chlorosis, frond disconnection, necrosis and colony disintegration (Greenberg et al., 1992; Prasad et al., 2001; Li and Xiong, 2004; Verma and Singh, 2006).

These changes in the associated transport host surface integrity and accumulated high levels of pollutants within tissues may interfere with the physical and chemical settlement cues used by free-living metacercariae as well as their ability to form a firm attachment to the host substrate. Thus, during the dispersal phase pollution may affect the establishment of these parasites in four basic ways: pollution of the water may kill or physiologically impair swimming cercariae inhibiting their ability to encyst; pollutants coating the surface of transport hosts may repel cercarial settlement either by killing the parasite on contact with the surface or interfering with settlement cues resulting in cercariae encysting elsewhere; pollution may change the surface properties of the transport host resulting in encysted metacercariae being unable to achieve a firm contact with the surface resulting in their easy dislodgement during host movements or in flowing water conditions (Figure 2).

8. CONCLUDING REMARKS

Free-living metacercariae demonstrate a sophisticated level of cercarial searching and settlement behaviour, comparable to that found for species with penetrative cercariae. It is evident that, despite the wide range of species involved and the diverse transport hosts utilized, they demonstrate many common characteristics in the mechanisms of location and levels of interactions they have with their hosts. These shared traits appear to be associated with the related ecology and behaviour of this life history rather than phylogenetic attributes, closely resembling the general biology demonstrated by disparate lecithotropic marine invertebrate larval species. The metacercarial—transport host relationship is analogous to that occurring between epibiotic organisms and their basibiont, with the same complex network of advantages and disadvantages that both may experience from the association. However, the relative importance of any given positive or negative interaction is likely dependent on a multitude of species-specific and environmental factors.

Nevertheless, it is apparent that some caution needs to be taken when extrapolating existing laboratory results to the natural environment, particularly associated with the duration of the dispersive phase and transport host

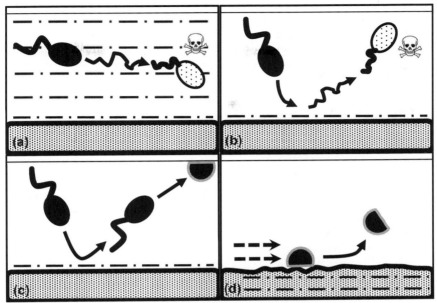

Figure 2 Potential effects of pollution on the establishment of free-living metacercarial populations: (a) pollution in water may kill/physiological impair swimming cercariae; (b) pollutants coating transport host surface may repel cercarial settlement by killing parasite; (c) pollutants coating transport host surface may interfere with settlement cues causing cercariae to encyst elsewhere; (d) high concentrations of pollutants in host tissue may change their outer surface properties resulting in the parasite unable to establish a firm contact causing dislodgement (— · — · — · — · occurrence of pollutants).

selection due to the constraints of the experimental approach when determining cercarial behaviour. The limitations of laboratory investigations of marine invertebrate larval behaviour are well known (Sulkin, 1986, 1990; Forward, 1988), particularly the assumption that simply miniaturizing the natural water column, whether it is a metre-high observation chamber or a simple depression slide, produces results that are directly transferable to the real world. Only by achieving a correct balance between laboratory and field studies is a proper understanding of the ecology of these parasites likely to be achieved.

However, in comparison to species that reside within an intermediate host, free-living metacercariae in many aspects are poorly understood. This review has highlighted much that is intriguing regarding both their pre-settlement and settlement phases that require further detailed elucidation, with many potentially productive avenues of investigation yet to be explored, particularly the nature of their relationship with the transport host. It is hoped that this review will provide the springboard for a greater level of attention to now be focused on this fascinating group of trematodes.

REFERENCES

Abbott, L.L., Bergey, E.A., 2007. Why are there few algae on snail shells? the effects of grazing, nutrients and shell chemistry on the algae on shells of *Helisoma trivolvis*. Freshwater Biol. 52, 2112—2120.

Abrahams, P.W., 2005. Geophagy and the involuntary ingestion of soil. In: Selinus, O. (Ed.), Essentials of Medical Geology. Elsevier, pp. 435—458.

Abrous, M., Vareille-Morel, C., Rondelaud, D., Dreyfuss, G., Cabaret, J., 2001. Metacercarial aggregation in Digenea (*Fasciola hepatica* and *Paramphistomum daubneyi*): environmental or species determinism? J. Helminthol. 75, 307—311.

Alda, P., Martorelli, S.R., 2014. Larval trematodes infecting the South-American intertidal mud snail *Heleobia australis* (Rissooidea: Cochliopidae). Acta Parasitol. 59, 50—67.

Alexeyev, V.M., 1962. The role of shrimps in the distribution of Notocotylosis. Zool. Zhurnal 41, 1255—1257 (In Russian).

Al-Jahdali, M.O., El-Said Hassanine, R.M., 2012. The life cycle of *Gyliauchen volubilis* Nagaty, 1956 (Digenea: Gyliauchenidae) from the Red Sea. J. Helminthol. 86, 165—172.

Almeida, M.J., Machado, J., Moura, G., Azevedo, M., Coimbra, J., 1998. Temporal and local variations in biochemical composition of *Crssostrea gigas* shells. J. Sea Res. 40, 233—249.

Angel, L.M., 1954. *Parorchis acanthus* var. *australis*, n. var., with an account of the life cycle in South Australia. Trans. Roy. Soc. S. Aust. 77, 164—174.

Asanji, M.F., Williams, M.O., 1974. Studies on the excystment of trematode metacercariae in vivo. J. Helminthol. 48, 85—91.

Asanji, M.F., Williams, M.O., 1985. Effect of age and environmental factors on the viability and excystment of metacercarial cysts of *Parorchis acanthus* in vivo and in vitro. Z. Parasitenkd. 71, 595—601.

Li, T.Y., Xiong, Z.T., 2004. Cadmium-induced colony disintegration of duckweed (*Lemna paucicostata* Hegelm.) and as a biomarker of phytotoxicity. Ecotoxicol. Environ. Saf. 59, 174—179.

Barlow, C.H., 1923. Life cycle of *Fasciolopsis buski* (human) in China. China Med. J. 37, 453—472.

Bauer, R.T., 1981. Grooming behavior and morphology in the decapod Crustacea. J. Crust. Biol. 1, 153—173.

Bennett, H.J., 1936. The life history of *Cotylophoron cotylophorum*, a trematode of ruminants. Ill. Biol. Monogr. 14, 1—119.

Bennett, C.E., 2001. *Fasciola hepatica*: surfaces involved in movement of miracidia and cercariae. J. Helminthol. 75, 1—5.

Bennett, S.C., Irwin, S.W.B., Fitzpatrick, S.M., 2003. Tributyltin and copper effects on encystment and in vitro excystment of *Parorchis acanthus* larvae. J. Helminthol. 77, 291—296.

Besprozvannykh, V.V., 2010. Lifecycle of the trematode *Notocotylus intestinalis* (Digenea, Notocotylidae) under natural conditions in Primorye region (Russia). Vestn. Zool. 44, 261—264.

Beuret, J., Pearson, J.C., 1994. Description of a new zygocercous cercaria (Opisthorchioidea: Heterophyidae) from prosobranch gastropods collected at Heron Island (Great Barrier Reef, Australia) and a review of zygocercariae. Syst. Parasitol. 27, 105—125.

Biermann, C.H., Schinner, G.O., Strathmann, R.R., 1992. Influence of solar radiation, microalgal fouling, and current on deposition site and survival of embryos of a dorid nubranch gastropod. Mar. Ecol. Prog. Ser. 86, 205—215.
Bolek, M.G., Janovy Jr., J., 2008. Alternative life cycle strategies of *Megalodiscus temperatus* in tadpoles and metamorphosed anurans. Parasite 15, 396—401.
Boray, J.C., 1963. The ecology of *Fasciola hepatica* with particular reference to its intermediate host in Australia. Proc. World Vet. Congr. 17, 709—715.
Boray, J.C., 1969. Experimental fascioliasis in Australia. Adv. Parasitol. 7, 95—210.
Boray, J.C., Enigk, K., 1964. Laboratory studies on the survival and infectivity of *Fasciola hepatica-* and *F. gigantica-* metacercariae. Z. Tropenmed. Parasitol. 15, 324—331.
Borowitzka, M.A., Lethbridge, R.C., Charlton, L., 1990. Species richness, spatial-distribution and colonization pattern of algal and invertebrate epiphytes on the seagrass *Amphibolis griffithii*. Mar. Ecol. Prog. Ser. 64, 281—291.
Bouix-Busson, D., Rondelaud, D., Combes, C., 1985. L'infestation de jeunes *Lymnaea glabra* Muller par *Fasciola hepatica* L. Les caracteristiques des emissions cercariennes. Ann. Parasitol. Hum. Comp. 60, 11—21.
Brglez, J., Wikerhauser, T., 1968. On the effect of some fertilizers and copper sulphate upon the survival of cercariae and metacercariae of *Fasciola hepatica*. Wiad. Parazytol. 14, 675—677.
Bulthuis, D.A., Woelkerling, W.J., 1983. Biomass accumulation and shading effects of epiphytes on leaves of seagrass, *Heterozostera tasmanica*, in Victoria, Australia. Aquat. Bot. 16, 137—148.
Burgu, A., 1982. Kistlenme sirasinda *Paramphistomum cervi* serkerlerinde renk secimi. Ankara Univ. Vet. Fak. 29, 143—150.
Burns, W.C., 1961. The life history of *Sphaeridiotrema spinoacetabulum* sp. n. (Trematoda: Psilostomidae) from the ceca of ducks. J. Parasitol. 47, 933—938.
Cancino, J.M., Munoz, J., Munoz, M., Orellana, M.C., 1987. Effects of the bryozoan *Membranipora tuberculata* (Bosc) on the photosynthesis and growth of *Gelidium rex* Santelices et Abbott. J. Exp. Mar. Biol. Ecol. 113, 105—112.
Casola, E., Scardi, M., Mazella, L., Fresi, E., 1987. Structure of the epiphytic community of *Posidonia oceanica* leaves in a shallow meadow. Mar. Ecol. 8, 286—296.
Chadhri, S.S., Gupta, R.P., 1985. Viability and infectivity of *Paramphistomum* metacercariae stored under different conditions. Indian Vet. J. 62, 470—472.
Cheng, T.C., Thakur, A.S., 1967. Thermal activation and inactivation of *Philophthalmus gralli*. J. Parasitol. 53, 212—213.
Cheruiyot, H.K., Wamae, L.W., 1990. Distribution of metacercariae of *Fasciola gigantica* on some objects in the laboratory. Bull. Anim. Health Prod. Afr. 38, 139—142.
Chowaniec, W., Markiewicz, K., 1970. A study on the virulence of *Fasciola hepatica* metacercariae. Acta Parasitol. Pol. 18, 25—32.
Christensen, N.O., Frandsen, F., Roushdy, M.Z., 1980. The influence of environmental conditions and parasite-intermediate host-related factors on the transmission of *Echinostoma liei*. Z. Parasitenkd. 63, 47—63.
Chubb, J.C., 1979. Seasonal occurence of helminths in freshwater fishes. Part II. Trematoda. Adv. Parasitol. 17, 141—313.
Claugher, D., 1976. A trematode associated with the death of the white-faced storm petrel (*Pelagodroma marina*) on the Chatham Islands. J. Nat. Hist. 10, 633—641.
Coil, W.H., 1984. An analysis of swimming by the cercariae of *Fasciola hepatica* using high speed cinematography. Proc. Helminthol. Soc. Wash. 51, 293—296.
Combes, C., Fournier, A., Mone, H., Theron, A., 1994. Behaviours in trematode cercariae that enhance parasite transmission: patterns and processes. Parasitology 109, S3—S13.
da Costa, C., Dreyfuss, G., Rakotondravao, Rodelaud, D., 1994. Several observations concerning cercarial sheddings of *Fasciola gigantica* from *Lymnaea natalensis*. Parasite 1, 39—44.

Cremonte, F., Kroeck, M.A., Martorelli, S.R., 2001. A new monorchiid cercaria (Digenea) parasitizing the purple clam *Amiantis purpurata* (Bivalvia: Veneridae) in the southwest Atlantic Ocean, with notes on its gonadal effect. Folia Parasitol. 48, 217—223.

Crisp, D.J., 1974. Factors influencing the settlement of marine invertebrate larvae. In: Grant, P.T., Mackie, A.M. (Eds.), Chemoreception in Marine Organisms. Academic Press, London, pp. 177—265.

Crisp, D.J., 1976. Settlement responses in marine organisms. In: Newell, R.C. (Ed.), Adaptation to Environment. Butterworths, London, pp. 83—124.

Cunningham, P.A., 1976. Inhibition of shell growth in the presence of mercury and subsequent recovery of juvenile oysters. Proc. Natl. Shellfish Ass. 66, 1—5.

Dalton, J.P. (Ed.), 1999. Fasciolosis. CABI Publishing, Wallingford.

Davtyan, E.A., 1956. Pathogenicity of different species of *Fasciola* and its variability depending on the developmental conditions of the parthenogenetic stages. Zool. Zhurnal 35, 1617—1625 (In Russian).

Diaz Diaz, M.T., 1976. Studies on Life-Cycles of Digenetic Trematodes (Ph.D. thesis). University of Leeds, UK.

Diaz, M.T., 1980. Ciclo vital de *Sphaeridiotrema newmillerdamensis* n.sp. (Platyhelminthes, Trematoda). Acta Biol. Venez. 10, 497—521.

Diaz, M.T., Hernandez, L.E., Bashirullah, A.K., 2002. Experimental life cycle of *Philophthalmus gralli* (Trematoda: Philophthalmidae) in Venezuela. Rev. Biol. Trop. 50, 629—642.

Diaz, M.T., Bashirullah, A.K., Hernandez, L.E., Gomez, E., 2009. Life cycle of *Culuwiya tilapiae* (Nasir y Gomez, 1976) (Trematoda: Haploporidae) in Venezuela. Rev. Cient. 19, 439—445.

Dixon, K.E., 1965. The structure and histochemistry of the cyst wall of the metacercaria of *Fasciola hepatica* L. Parasitology 55, 215—226.

Dixon, K.E., 1966. The physiology of excystment of the metacercaria of *Fasciola hepatica* L. Parasitology 56, 431—456.

Dixon, K.E., Mercer, E.H., 1967. The formation of the cyst wall of the metacercaria of *Fasciola hepatica* L. Z. Zellforsch. Microsk. Anat. 77, 345—360.

Dobretsov, S., Abed, R.M., Voolstra, C.R., 2013. The effect of surface colour on the formation of marine micro and macrofouling communities. Biofouling 29, 617—627.

Dollfus, R.P., 1966. Métacercaire énigmatique de distome, du plankton de surface des Iles du Cap Vert. Bull. Mus. Natl. Hist. Nat. 38, 195—200.

Dönges, J., 1962. Entwicklungsgeschichtliche und morphologische untersuchungen an Notocotyliden (Trematoda). Z. Parasitenkd. 22, 43—67.

Dönges, J., 1969. Entwicklungs- und Lebensdauer von Metacercarien. Z. Parasitenkd. 31, 340—366.

Dreyfuss, G., Rondlelaud, D., 1994. *Fasciola hepatica*: a study of the shedding of cercaraie from *Lymnaea truncatula* raised under constant conditions of temperature and photoperiod. Parasite 1, 401—404.

Dreyfuss, G., Dacosta, C., Rakotondravao, Rondelaud, D., 1995. *Fasciola gigantica*- the parasite burden in *Lymnaea natalensis* that died after a cercarial shedding. Parasite 2, 177—180.

Dreyfuss, G., Abrous, M., Vignoles, P., Rondelaud, D., 2004. *Fasciola hepatica* and *Paramphistomum daubneyi*: vertical distribution of metacercariae on plants under natural conditions. Parasitol. Res. 94, 70—73.

Dreyfuss, G., Vignoles, P., Rondelaud, D., 2005. *Fasciola hepatica*: epidemiological surveillance of natural watercress beds in central France. Parasitol. Res. 95, 278—282.

Dreyfuss, G., Vignoles, V., Rondelaud, D., 2009. The redial and cercarial production of a digenean in the snail host is lower when no cercarial shedding occurs. Parasite 16, 309—313.

Durie, P.H., 1955. A technique for the collection of large numbers of paramphistome (trematoda) metacercariae. Aust. J. Agric. Res. 6, 200—202.

Durie, P.H., 1956. The paramphistomes (Trematoda) of Australian ruminants. III. The life-history of *Calicophoron calicophorum* (Fischoeder) Nasmark. Aust. J. Zool. 4, 152—157.

Dutt, S.C., Srivastava, H.D., 1972. The life history of *Gastrodiscoides hominis* (Lewis and McConnel, 1876) Leiper, 1913- the amphistome parasite of man and pig. J. Helminthol. 46, 35—46.

Dyrynda, P.E.J., 1986. Defensive strategies of modular organisms. Phil. Trans. R. Soc. B 313, 227—243.

Efford, I.E., Tsumura, K., 1969. Observations on the biology of the trematode *Megalodiscus microphagus* in amphibeans from Marion Lake, British Columbia. Am. Midl. Nat. 82, 197—203.

Eldred, B., 1962. The attachment of the barnacle, *Balanus amphitrite niveus* Darwin, and other fouling organisms to the rock shrimp, *Sicyonia dorsalis* Kingsley. Crustaceana 3, 203—206.

El-Mayass, H., 1991. A Study of Digenean Parasites from the Salt Marshes of North Norfolk (Ph.D. thesis). University of East Anglia, UK.

Erasmus, D.A., 1972. The Biology of Trematodes. Edward Arnold, London.

Erkina, N.G., 1954. The life-cycle of the trematode *Notocotylus chionis*, parasite of aquatic birds. Dokl. Akad. Nauk. SSSR 97, 559—560 (In Russian).

Esclaire, F., Audousset, J.C., Rondelaud, D., Dreyfuss, G., 1989. Les metacercaires 'flottantes' de *Fasciola hepatica* L. a propos de quelques observations sur leur structure et leurs variations numeriques au cours d'une infestation experimentale chez *Lymnaea truncatula* Muller. Bull. Soc. Franc. Parasitol. 7, 225—228.

Esteban, J.G., Gonzalez, C., Bargues, M.D., Angles, R., Sanchez, C., Naquira, C., Mas-Coma, S., 2002. High fascioliasis infection in children linked to a man-made irrigation zone in Peru. Trop. Med. Int. Health 7, 339—348.

Evans, N.A., 1982. Effects of copper and zinc on the life cycle of *Notocotylus attenuatus* (Digenea: Notocotylidae). Int. J. Parasitol. 12, 363—369.

Evans, D.W., Irwin, S.W.B., Fitzpatrick, S.M., 1997. Metacercarial encystment and *in vivo* cultivation of *Cercaria lebouri* Stunkard 1932 (Digenea: Notocotylidae) to adult identified as *Paramonostomum chabaudi* Van Strydonck 1965. Int. J. Parasitol. 27, 1299—1304.

Ferrell, D.L., Negovetich, N.J., Wetzel, E.J., 2001. Effect of temperature on the infectivity of metacercariae of *Zygocotyle lunata* (Digenea: Paramphistomidae). J. Parasitol. 87, 10—13.

Fingerut, J.T., Zimmer, C.A., Zimmer, R.K., 2003. Larval swimming overpowers turbulent mixing and facilitates transmission of a marine parasite. Ecology 84, 2502—2515.

Forward Jr., R.B., 1988. Diel vertical migration: zooplankton photobiology and behaviour. Oceanogr. Mar. Biol. Annu. Rev. 26, 361—393.

Fried, B., 1981. Thermal activation and inactivation of the metacercariae of *Philophthalmus hegeneri* Penner and Fried 1963. Z. Parasitenkd. 65, 359—360.

Fried, B., 1994. Metacercarial excytment of Trematodes. Adv. Parasitol. 33, 91—144.

Fried, B., Weaver, L.J., 1969. Exposure of chicks to the metacercaria of *Echinostoma revolutum* (Trematoda). Proc. Helminthol. Soc. Wash. 36, 153—155.

Fried, B., Grigo, K.L., 1975. Encystment of *Philophthalmus hegeneri* (Trematoda) cercariae on *Artemia salina* (Crustacea). Proc. Helminthol. Soc. Wash. 42, 63—65.

Fried, B., Bennett, M.C., 1979. Studies on encystment of *Echinostoma revolutum* cercariae. J. Parasitol. 65, 38—40.

Fried, B., Robbins, S.H., Nelson, P.D., 1978. In vivo and in vitro excystation of *Zygocotyle lunata* (Trematoda) metacercariae and histochemical observations on the cyst. J. Parasitol. 64, 395—397.

Fried, B., Schmidt, K.A., Sorensen, R.E., 1997. In vivo and ectopic encystment of *Echinostoma revolutum* and chemical excystation of the metacercariae. J. Parasitol. 83, 251—254.

Fried, B., Peoples, R.C., Saxton, T.M., Huffman, J.E., 2008. The association of *Zygocotyle lunata* and *Echinostoma trivolvis* with *Chaetogaster limnaei*, an ectosymbiont of *Helisoma trivolvis*. J. Parasitol. 94, 553—554.

Fried, B., Huffman, J.E., Keeler, S., Peoples, R.C., 2009. The biology of the caecal trematode *Zygocotyle lunata*. Adv. Parasitol. 69, 1–40.
Gagne, F., Blaise, C., Pellerin, J., 2005. Altered exoskeleton composition and vitellogenesis in the crustacean *Gammarus* sp. collected at polluted sites in the Saguenay Fjord, Quebec, Canada. Environ. Res. 98, 89–99.
Galaktionov, K., Dobrovolskij, A., 2003. The Biology and Evolution of Trematodes: An Essay on the Biology, Morphology, Life Cycles, Transmissions, and Evolution of Digenetic Trematodes. Kluwer, Dordrecht.
Gibson, D.I., Bray, R.A., 1977. The Azygiidae, Hirudinellidae, Ptychogonimidae, Sclerodistomidae and Syncoeliidae (Digenea) of fishes from the northeast Atlantic. Bull. Br. Mus. Nat. Hist. Zool. 32, 167–245.
Ginetsinskaya, T.A., 1988. Trematodes, Their Life Cycles, Biology and Evolution. Amerind Publishing Company, New Delhi.
Glynn, P.W., 1970. Growth of algal epiphytes on a tropical marine isopod. J. Exp. Mar. Biol. Ecol. 5, 88–93.
Grabda-Kazubska, B., 1980. The life cycle of *Diplodiscus subclavatus* (Trematoda, Diplodiscidae). Acta Parasitol. Pol. 27, 261–272.
Graczyk, T.K., Shiff, C.J., 1994. Viability of *Notocotylus attenuatus* (Trematoda: Notocotylidae) metacercariae under adverse conditions. J. Wildl. Dis. 30, 46–50.
Graefe, G., 1971. Ungewöhnliches Verhalten einer argentinischen Haploporidencercarie. Parasitol. Schriftenr. 21, 179–182.
Green, P., Dussart, G.B.J., Gibson, C., 1992. Surfacing and water leaving behaviour of the freshwater pulmonate snails *Lymnaea peregra* (Muller), *Biomphalaria glabrata* (Say) and *Bulinus jousseaumei* (Dautzenberg). J. Mollusc. Stud. 58, 169–179.
Greenberg, B.M., Huang, X.-D., Dixon, D.G., 1992. Applications of the aquatic higher plant *Lemna gibba* for ecotoxicological assessment. J. Aquat. Ecosyst. Health 1, 147–155.
Griffiths, H.J., Christensen, C.A., 1972. Survival of metacercariae of *Fascioloides magna* in water at room temperature and under refrigeration. J. Parasitol. 58, 404–405.
Griffiths, H.J., Christensen, C.A., 1974. Further observations on the survival of metacercariae of *Fascioloides magna* in water at room temperature and under refrigeration. J. Parasitol. 60, 335.
Gupta, S.C., Kamra, D.N., 1987. Influence of wastelage fermentation on viability of *Fasciola gigantica* metacercariae. Biol. Wastes 22, 311–313.
Gutiérrez, J.L., Jones, C.G., Strayer, D.L., Iribarne, O.O., 2003. Mollusks as ecosystem engineers: the role of shell production in aquatic habitats. Oikos 101, 79–90.
Haas, W., 1992. Physiological analysis of cercarial behavior. J. Parasitol. 78, 243–255.
Haas, W., 1994. Physiological analyses of host-finding behaviour in trematode cercariae: adaptations for transmission success. Parasitology 109, S15–S29.
Haas, W., 2003. Parasitic worms: strategies of host finding, recognition and invasion. Zoology 106, 349–364.
Hadfield, M.G., Paul, V.J., 2001. Natural chemical cues for settlement and metamorphosis of marine-invertebrate larvae. In: McClintock, J.B., Baker, B.J. (Eds.), Marine Chemical Ecology. CRC Press, pp. 431–462.
Hammond, J.A., 1974. Human infection with the liver fluke *Fasciola gigantica*. Trans. R. Soc. Trop. Med. Hyg. 68, 253–254.
Harper, W.F., 1929. On the structure and life histories of British fresh-water larval trematodes. Parasitology 21, 189–219.
Harris, A.L., 1986. Larval Trematode Infections of the Freshwater Snail *Lymnaea Peregra* (Muller) (M.Phil. thesis). Queen Mary & Westfield College, University of London, UK.
Herrmann, K.K., Sorensen, R.E., 2009. Seasonal dynamics of two mortality-related trematodes using an introduced snail. J. Parasitol. 95, 823–828.

Hidalgo, F.J., Firstater, F.N., Lomovasky, B.J., Gallegos, P., Gamero, P., Iribarne, O.O., 2010. Microalgal fouling on the intertidal mole crab *Emerita analoga* facilitates bird predation. Helgol. Mar. Res. 64, 367–376.

Hodasi, J.K.M., 1972. The output of cercariae of *Fasciola hepatica* by *Lymnaea truncatula* and the distribution of metacercariae on grass. Parasitology 64, 53–60.

Hohman, W.L., 1985. Feeding ecology of ringed-neck ducks in Northwestern Minesota. J. Wildl. Manage. 49, 546–557.

Holliman, R.B., 1961. Larval trematodes from the Apalachee bay area, Florida, with a checklist of known marine cercariae arranged in a key to their superfamilies. Tulane Stud. Zool. 9, 1–74.

Honer, M.R., 1961. *Diplodiscus subclavatus* (Goeze 1782) var. *paludinae*, var. nov., from *Paludina vivipara* Lm. in the Netherlands. Z. Parasitenkd. 20, 489–494.

Hoover, R.C., Lincoln, S.D., Hall, R.F., Wescott, R., 1984. Seasonal transmission of *Fasciola hepatica* to cattle in northwestern United States. J. Am. Vet. Med. Assoc. 184, 695–698.

Howell, M.J., 1983. Gland cells and encystment of *Philopthalmus burrili* cercariae. Z. Parasitenkd. 69, 207–216.

Howell, M.J., Bearup, A.J., 1967. The life histories of two bird trematodes of the family Philophthalmidae. Proc. Linn. Soc. N.S.W. 92, 182–194.

Huffman, J.E., 1986. Structure and composition of the metacercarial cyst wall of *Sphaeridiotrema globulus* (Trematoda). Int. J. Parasitol. 16, 647–653.

Hunter, W.S., 1967. Notes on the life history of *Pleurogonius malaclemys* Hunter, 1961 (Trematoda: Pronocephalidae) from Beaufort, North Carolina, with a description of the cercaria. Proc. Helminthol. Soc. Wash. 34, 33–40.

Humiczewska, M., 2004. Some enzymes of respiratory chain in metacercariae of *Fasciola hepatica*. Zool. Pol. 49, 63–76.

Huspeni, T.C., Lafferty, K.D., 2004. Using larval trematodes that parasitize snails to evaluate a saltmarsh restoration project. Ecol. Arch. A014-016-A1.

Imber, M.J., 1984. Trematode anklets on whitefaced stormpetrels *Pelagodroma marina* and fairy prions *Pachyptila turtur*. Cormorant 12, 71–74.

Irwin, S.W.B., 1997. Excystation and cultivation of trematodes. In: Fried, B., Graczyk, T.K. (Eds.), Advances in Trematode Biology. CRC Press, Boca Raton, pp. 57–86.

James, B.L., 1971. Host selection and ecology of marine digenean larvae. In: Crisp, D.J. (Ed.), Fourth European Marine Biological Symposium. Cambridge University Press, Cambridge, pp. 179–196.

Jimenez-Albarran, M., Guevara-Pozo, D., 1980. Estudios experimentales sobre biologia de *Fasciola hepatica*: 5- Influencia del color de la luz en la fijacion de las metacercarias sobre la superficie del vidrio. Rev. Iber. Parasitol. 40, 443–452.

Jousson, O., Bartoli, P., 1999. The life-cycle of three species of the Mesometridae (Digenea) with comments on the taxonomic status of this family. Syst. Parasitol. 44, 217–228.

Kádár, E., Costa, V., 2006. First report on the micro-essential metal concentrations in bivoalve shells from deep-sea hydrothermal vents. J. Sea Res. 56, 37–44.

Keeler, S.P., Huffman, J.E., 2009. Echinostomes in the second intermediate host. In: Fried, B., Toledo, R. (Eds.), The Biology of Echinostomes. Springer, New York, pp. 61–87.

Kendall, S.B., McCullough, F.S., 1951. The emergence of *Fasciola hepatica* from the snail *Limnaea truncatula*. J. Helminthol. 25, 77–92.

Koehler, A.V., Brown, B., Poulin, R., Thieltges, D.W., Fredensborg, B.L., 2012. Disentangling phylogenetic constraints from selective forces in the evolution of trematode transmission stages. Evol. Ecol. 26, 1497–1512.

Komiya, Y., 1964. Fasciolopsis buski. Prog. Med. Parasitol. Jap. 1, 277–285.

Komiya, Y., Suzuki, N., 1964a. Biology of *Clonorchis sinensis*. Prog. Med. Parasitol. Jap. 1, 553–600.

Komiya, Y., Suzuki, N., 1964b. The distribution of *Clonorchis* infection in Japan with remarks on its epidemiology. Prog. Med. Parasitol. Jap. 1, 603—660.
Koprivnikar, J., Poulin, R., 2009. Effects of temperature, salinity, and water level on the emergence of marine cercariae. Parasitol. Res. 105, 957—965.
Koprivnikar, J., Lim, D., Fu, C., Brack, S.H.M., 2010. Effects of temperature, salinity, and pH on the survival and activity of marine cercariae. Parasitol. Res. 106, 1167—1177.
Kouchi, N., Nakaoka, M., Mukai, H., 2006. Effects of temporal dynamics and vertical structure of the seagrass *Zostera caulescens* on distribution and recruitment of the epifaunal encrusting bryozoan *Microporella trigonellata*. Mar. Ecol. 27, 145—153.
Krug, P.J., 2006. Defence of benthic invertebrates against surface colonization by larvae: a chemical arms race. Progr. Mol. Subcell. Biol. 42, 1—53.
Krull, W.H., 1934. Life history studies on *Cotylophoron cotylophorum* (Fischoeder, 1901) Stiles and Goldberger, 1910. J. Parasitol. 20, 173—180.
Krull, W.H., Price, H.F., 1932. Studies on the life history of *Diplodiscus temperatus* Stafford from the frog. Occas. Pap. Mus. Zool. Univ. Mich. 237, 1—38.
Kyle, D.E., Noblet, G.P., 1985. Occurrence of metacercariae (Trematoda: Gymnophallidae) on *Amphitrite ornate* (Annelida: Terebellidae). J. Parasitol. 71, 366—368.
Lang, A., 1892. Ueber die cercarie von *Amphistomum subclavatum*. Ber. Naturforsch. Ges. Freib. I. Br. 6, 81—89.
Laurie, J.S., 1974. *Himasthla quissetensis*: induced in vitro encystment of cercaria and ultrastructure of the cyst. Exp. Parasitol. 35, 350—362.
Lei, F., Poulin, R., 2011. Effects of salinity on multiplication and transmission of an intertidal trematode parasite. Mar. Biol. 158, 995—1003.
Lepitzki, D.A.W., 1993. Epizootiology and Transmission of Snail-Inhabiting Metacercariae of the Duck Digeneans *Cyathocotyle bushiensis* and *Sphaeridiotrema globulus* (Ph.D. thesis). McGill University, Canada.
Lepitzki, D.A.W., Bunn, B.M., 1994. A plug in the cyst wall of metacercariae of *Sphaeridiotrema pseudoglobulus* (Digenea: Psilostomidae) and a possible novel mode of transmission. Int. J. Parasitol. 24, 273—275.
Lepitzki, D.A.W., Scott, M.E., McLaughlin, J.D., 1994. Assessing cercarial transmission of *Cyathocotyle bushiensis* and *Sphaeridiotrema pseudoglobulus* by use of sentinel snails. Can. J. Zool. 72, 885—891.
LeSage, K.E., Fried, B., 2011. Encystment and excystment of the paramphistomid trematode *Zygocotyle lunata*. J. Helminthol. 83, 300—303.
Lewtas, K.L.M., Birch, G.F., Foster-Thorpe, C., 2014. Metal accumulation in the greentail prawn, *Metapenaeus bennettae*, in Sydney and Port Hacking estuaries, Australia. Environ. Sci. Pollut. Res. 21, 704—716.
Loh, K., Kim, J.J., Hyun, J.K., Namgoong, T., 1969. Experimental observations on water contamination by the second intermediate host infected with *Paragonimus westermani*. Korean J. Parasitol. 7, 1—5 (In Korean).
Looss, A., 1892. Ueber *Amphistomum subclavatum* Rud. und seine Entwicklung. In: Festschrift zum Siebenzigsten Geburtstages Rudolf Leuckart's; dem verehrten Jubilar dargebracht von seinen dankbaren Schülern. Wilhelm Englemann, Leipzig, pp. 147—167.
Lutz, A., 1892. Zur lebesgeschichte des *Distoma hepaticum*. Zentralblatt Bakteriol. Parasitenkd. 11, 783—796 (English translation in: Benchimol, J.L., Sá, M.R. (Eds.), Adolpho Lutz: Helmintologia, v.3, Livro 2, Editora Fiocruz, Rio de Janeiro, 2007, pp. 415—425.).
Luzón-Peña, L., Rojo-Vázquez, F.A., Gómez-Bautista, M., 1994. The overwintering of eggs, intramolluscan stages and metacercariae of *Fasciola hepatica* under the temperatures of a Mediterranean area (Madrid, Spain). Vet. Parasitol. 55, 143—148.
Macko, J.K., 1980. On some host problems from the viewpoint of species aspects. I. On helminth hosts in general and their terminology in homoxenous helminths (geohelminths). Helminthologia 17, 133—140.

Macy, R.W., Bell, W.D., 1968. The life cycle of *Astacatrematula macrocotyla* gen. et sp. n. (Trematoda: Psilostomidae) from Oregon. J. Parasitol. 54, 319—323.

Macy, R.W., Berntzen, A.K., Benz, M., 1968. In vitro excystation of *Sphaeridiotrema globulus* metacercariae, structure of cyst, and the relationship to host specificity. J. Parasitol. 54, 28—38.

McCarthy, A.M., 1999. Phototactic responses of the cercariae of *Echinoparyphium recurvatum* during phases of sub-maximal and maximal infectivity. J. Helminthol. 73, 63—65.

McEdward, L. (Ed.), 1995. Ecology of Marine Invertebrate Larvae. CRC Press, Boca Raton.

McMahon, R.F., 1990. Thermal tolerance, evaporative water loss, air-water oxygen consumption and zonation of intertidal prosobranchs: a new synthesis. Hydrobiologia 193, 241—260.

Martin, W.E., 1973. Life history of *Saccocoelioides pearsoni* n. sp. and the description of *Lecithobotrys sprenti* n. sp. (Trematoda: Haploporidae). Trans. Am. Microsc. Soc. 92, 80—95.

Mas-Coma, S., 2004. Human Fascioliasis. Waterborne Zoonoses: Identification, Causes and Control. World Health Organization (WHO)/IWA Publishing, London, pp. 305—322.

Meadows, P.S., Campbell, J.I., 1972. Habitat selection by aquatic invertebrates. Adv. Mar. Biol. 10, 271—382.

Moret, Y., Moreau, J., 2012. The immune role of the arthropod exoskeleton. Invert. Surviv. J. 9, 200—206.

Morley, N.J., Crane, M., Lewis, J.W., 2001. Toxicity of cadmium and zinc to encystment and *in vitro* excystment of *Parorchis acanthus* (Digenea: Philopthalmidae). Parasitology 122, 75—79.

Morley, N.J., Crane, M., Lewis, J.W., 2002. Toxicity of cadmium and zinc to encystment of *Notocotylus attenuatus* (trematoda: Notocotylidae) cercariae. Ecotoxicol. Environ. Saf. 53, 129—133.

Morley, N.J., Leung, K.M.Y., Morritt, D., Crane, M., 2003a. Toxicity of anti-fouling biocides to *Parorchis acanthus* (Digena: Philophthalmidae) cercarial encystment. Dis. Aquat. Org. 54, 55—60.

Morley, N.J., Irwin, S.W.B., Lewis, J.W., 2003b. Pollutant toxicity to the transmission of larval digeneans through their molluscan intermediate hosts. Parasitology 126, S5—S26.

Morley, N.J., Crane, M., Lewis, J.W., 2003c. Cadmium toxicity and snail-digenean interactions in a population of *Lymnaea* spp. J. Helminthol. 77, 49—55.

Morley, N.J., Crane, M., Lewis, J.W., 2004. Influence of cadmium exposure on the incidence of first intermediate host encystment by *Echinoparyphium recurvatum* (Digenea: Echinostomatidae) cercariae in *Lymnaea peregra*. J. Helminthol. 78, 329—332.

Morley, N.J., 2012. Cercariae (Platyhelminthes: trematoda) as neglected components of zooplankton communities in freshwater habitats. Hydrobiologia 691, 7—19.

Morley, N.J., Lewis, J.W., 2013. Thermodynamics of cercarial development and emergence in trematodes. Parasitology 140, 1211—1224.

Morley, N.J., Lewis, J.W., 2015. Thermodynamics of trematode infectivity. Parasitology 142, 585—597.

Mouritsen, K.N., Bay, G.M., 2000. Fouling of gastropods: a role for parasites? Hydrobiologia 418, 243—246.

Murrell, K.D., 1965. Stages in the life cycle of *Wardius zibethicus* Barker, 1915. J. Parasitol. 51, 600—604.

Murty, A.S., 1966. Experimental demonstration of the life-cycle of *Philophthalmus* sp. (Trematoda: Philophthalmidae). Curr. Sci. 14, 366—367.

Murty, A.S., 1973. Life cycle of *Pseudodiplodiscoides pilai* (Trematoda: Diplodiscidae) from the gut of the apple snail, *Pila globosa* (Swainson). J. Parasitol. 59, 323—326.

Nadakal, A.M., 1960. Chemical nature of cercarial eye-spot and other tissue pigments. J. Parasitol. 46, 475—483.

Nagano, K., 1964. The prevention of *Clonorchis sinensis*. Prog. Med. Parasitol. Jap. 1, 725—738.
Neal, A.T., Poulin, R., 2012. Substratum preference of *Philophthalmus* sp. cercariae for cyst formation under natural and experimental conditions. J. Parasitol. 98, 293—298.
Nollen, P.M., Kanev, I., 1995. The taxonomy and biology of Philopthalmid eyeflukes. Adv. Parasitol. 36, 205—269.
Nollen, P.M., Leslie, J.F., Cain, G.D., MacNab, R.K., 1985. A comparison of Texan and Hawaiian strains of the avian eyefluke, *Philophthalmus gralli*, with a cautionary note on the importation of exotic animals. J. Parasitol. 71, 618—624.
Nuñez, J.D., Laitano, M.V., Cledón, M., 2012. An intertidal limpet species as a bioindicator: pollution effects reflected by shell characteristics. Ecol. Indic. 14, 178—183.
Odening, K., 1976. Conception and terminology of hosts in parasitology. Adv. Parasitol. 14, 1—93.
Ollerenshaw, C.B., 1959. The ecology of liver fluke (*Fasciola hepatica*). Vet. Rec. 71, 957—965.
Ollerenshaw, C.B., 1971. Some observations on the epidemiology of Fascioliasis in relation to the timing of molluscicide applications in the control of the disease. Vet. Rec. 88, 152—164.
Overstreet, R.M., 1970. A syncoeliid (Hemiuroidea Faust, 1929) metacercaria on a copepod from the Atlantic equatorial current. J. Parasitol. 56, 834—836.
Owens, L., 1987. A checklist of metazoan parasites from Natantia (excluding the crustacean parasites of the Caridea). J. Shellfish Res. 6, 117—124.
Pantelouris, E.M., 1965. The Common Liver Fluke. Pergamon Press, Oxford.
Pecheur, M., 1967. La cercaire de *Fasciola hepatica*. Le role de la couleur de la lumière et des plantes sur le choix de l'endroit de fixation. La cercaire est-elle infestante? Ann. Méd. Vét. 6, 349—355.
Peoples, R.C., Fried, B., 2008. The effects of various chemical and physical factors on encystment and excystment of *Zygocotyle lunata*. Parasitol. Res. 103, 899—904.
Pfukenyi, D.M., Mukaratirwa, S., Willingham, A.L., Monrad, J., 2005. Epidemiological studies of amphistome infections in cattle in the highveld and lowveld communal grazing areas of Zimbabwe. Onderstepoort J. Vet. Res. 72, 67—86.
Pfukenyi, D.M., Mukaratirwa, S., Willingham, A.L., Monrad, J., 2006. Epidemiological studies of *Fasciola gigantica* infections in cattle in the highveld and lowveld communal grazing areas of Zimbabwe. Onderstepoort J. Vet. Res. 73, 37—51.
Pike, A.W., 1968. Observations on the life-cycle of *Psilotrema oligoon* (Linstow, 1887) Odhner, 1913, and on the larval stages of two other psilostome trematodes. Parasitology 58, 171—183.
Pike, A.W., 1969. Observations on the life cycles of *Notocotylus triserialis* Diesing, 1839, and *N. imbricatus* (Looss, 1893) *sensu* Szidat, 1935. J. Helminthol. 63, 145—165.
Pike, A.W., Erasmus, D.A., 1967. The formation, structure and histochemistry of the metacercarial cyst of three species of digenetic trematodes. Parasitology 57, 683—694.
Plakas, S.M., Doerge, D.R., Turnipseed, S.B., 1999. Disposition and metabolism of malachite green and other therapeutic dyes in fish. In: Smith, D.J., Gingerich, W.H., BeconiBarker, M.G. (Eds.), Xenobiotics in Fish. Springer, New York, pp. 149—166.
Platt, T.R., Dowd, R.M., 2012. Age-related change in phototaxis by cercariae of *Echinostoma caproni* (Digenea: Echinostomatidae). Comp. Parasitol. 79, 1—4.
Porter, A., 1938. The larval trematoda found in certain South African Mollusca. Pub. South Afr. Inst. Med. Res. 62, 1—492.
Prasad, A., Ghosh, S., Bhatnagar, P.K., Santra, P.K., 1999. Studies on the viability of metacercariae of *Fasciola gigantica*. J. Helminthol. 73, 163—166.
Prasad, M.N.V., Malec, P., Waloszek, A., Bojko, M., Strzalka, K., 2001. Physiological responses of *Lemna trisulca* L. (duckweed) to cadmium and copper bioaccumulation. Plant Sci. 161, 881—889.

Prinz, K., Kelley, T.C., O'Riordan, R.M., Culloty, S.C., 2009. Non-host organisms affect transmission processes in two common trematode parasites of rocky shores. Mar. Biol. 156, 2303—2311.

Prinz, K., Kelley, T.C., O'Riordan, R.M., Culloty, S.C., 2011. Factors influencing cercarial emergence and settlement in the digenean trematode *Parorchis acanthus* (Philophthalmidae). J. Mar. Biol. Assoc. U.K. 91, 1673—1679.

Probert, A.J., 1965. Studies on larval trematodes infecting the freshwater molluscs of Llangorse lake, south Wales. Part II. The Gymnocephalous cercariae. J. Helminthol. 39, 53—66.

Prokofyev, V.V., 1994. Lie-in-wait type of behaviour of cercariae of some marine trematodes. Zool. Zhurnal 73, 13—20 (In Russian, English translation in Hydrobiol. J., 1995, 31(4), 25—33.).

Radlett, A.J., 1978. Studies on Some Digenea (Platyhelminthes) (Ph.D. thesis). University of Hull, UK.

Rajasekariah, G.R., Howell, M.J., 1978. The passage of cysts of *Fasciola hepatica* in the feces of the snail intermediate host. Proc. Helminthol. Soc. Wash. 45, 138—139.

Ray, S., Pandey, K.C., Agrawal, N. A new species of an interesting amphistome *Pseudodiplodiscoides* Murty, 1970 in *Bellamya bengalensis* Lamarck, 1882 from water bodies near district Barabanki, U.P. J. Parasit. Dis., in press.

Rea, J.G., Irwin, S.W.B., 1995. The effects of age, temperature and shadow stimuli on activity patterns of the cercariae of *Cryptocotyle lingua* (Digenea: Heterophyidae). Parasitology 111, 95—101.

Rees, G., 1937. The anatomy and encystment of *Cercaria purpurae* Lebour, 1911. Proc. Zool. Soc. Lond. B107, 65—73.

Rees, G., 1948. A study of the effect of light, temperature and salinity on the emergence of *Cercaria purpurae* Lebour from *Nucella lapillus* (L.). Parasitology 38, 228—242.

Rees, G., 1967. The histochemistry of the cytogenous gland cells and cyst wall of *Parorchis acanthus* Nicoll, and some details of the morphology and fine structure of the cercaria. Parasitology 57, 87—110.

Rees, G., 1971. Locomotion of the cercaria of *Parorchis acanthus*, Nicoll and the ultrastructure of the tail. Parasitology 62, 489—503.

Rizvi, A., Alam, Md.M., Parveen, S., Saleemuddin, M., Abidi, S.M.A., 2012. Abandoning the ship: spontaneous mass exodus of *Clinstomum complanatum* (Rudolphi, 1814) progenetic metacercariae from the dying host *Trichogaster fasciatus* (Bloch & Schneider, 1801). J. Parasit. Dis. 36, 139—140.

Robson, M.A., Williams, D., Wolff, K., Thomason, J.C., 2009. The effect of surface colour on the adhesion strength of *Elminius modestus* Darwin on a commercial non-biocidal antofouling coating at two locations in the UK. Biofouling 25, 215—217.

Rodriguez, S.R., Ojeda, F.P., Inestrosa, N.C., 1993. Settlement of benthic marine invertebrates. Mar. Ecol. Prog. Ser. 97, 193—207.

Rohde, S., Hiebenthal, C., Wahl, M., Karez, R., Bischof, K., 2008. Decreased depth distribution of *Fucus vesiculosus* (Phaeophyceae) in the Western Baltic: effects of light deficiency and epibionts on growth and photosynthesis. Eur. J. Phycol. 43, 143—150.

Rondelaud, D., 1993. Variabilite interpopulationnelle de l'infestation faciolienne chez le mollusque *Lymnaea truncatula* Muller. Influence du contact prealable de la population avec le parasite. Bull. Soc. Zool. Fr. 118, 185—193.

Rondelaud, D., Vignoles, P., Vareille-Morel, C., Abrous, M., Mage, C., Mouzet, R., Dreyfuss, G., 2004. *Fasciola hepatica* and *Paramphistomum daubneyi*: field observations on the transport and outcome of floating metacercariae in running water. J. Helminthol. 78, 173—177.

Rondelaud, D., Hourdin, P., Vignoles, P., Dreyfuss, G., 2005. The contsmination of wild watercress with *Fasciola hepatica* in central France depends on the ability of several lymnaeid snails to migrate upstream towards the beds. Parasitol. Res. 95, 305—309.

Rondelaud, D., Hourdin, P., Vignoles, P., Dreyfuss, G., Cabaret, J., 2011. The detection of snail host habitats in liver fluke infected farms by use of plant indicators. Vet. Parasitol. 181, 166—173.

Rondelaud, D., Titi, A., Vignoles, P., Mekroud, A., Dreyfuss, G., 2013. Consequence of temperature changes on cercarial shedding from *Galba truncatula* infected with *Fasciola hepatica* or *Paramphistomum daubneyi*. Parasite 20, 10.

Roque, F.T., Ludwick, R.W., Bell, J.C., 1953. Pulmonary Paragonimiasis: a review with case reports from Korea and the Philippines. Ann. Intern. Med. 38, 1206—1221.

Ross, J.G., 1977. 5-year study of epidemiology of Fascioliasis in N, E and W of Scotland. Brit. Vet. J. 133, 263—272.

Ryan, P.G., 1986. Trematode anklets on Procellariform seabirds from southern Africa and the adjacent southern ocean. Cormorant 13, 157—161.

Sand-Jensen, K., 1977. Effect of epiphytes on eelgrass photosynthesis. Aquat. Bot. 3, 55—63.

Samson, K.S., Wilson, G.I., 1974. Passage through Rouen Ducks of the metacercariae of *Fasciola hepatica*. Proc. Helminthol. Soc. Wash. 41, 112—113.

Satheesh, S., Godwin Wesley, S., 2010. Influence of substratum colour on the recruitment of macrofouling communities. J. Mar. Biol. Assoc. U.K. 90, 941—946.

Sey, O., 1983. Reconstruction of the systematics of the family Diplodiscidae Skrjabin, 1949 (Trematoda: Paramphistomata). Parasitol. Hung. 16, 63—89.

Shameem, U., Madhavi, R., 1991. Observations on the life-cycles of two haploporid trematodes, *Carassotrema bengalense* Rekharani & Madhavi, 1985 and *Saccocoelioides martini* Madhavi, 1979. Syst. Parasitol. 20, 97—107.

Shostak, A.W., Dharampaul, S., Belosevic, M., 1993. Effects of source of metacercariae on experimental infection of *Zygocotyle lunata* (Digenea: Paramphistomidaae) in CD-1 mice. J. Parasitol. 79, 922—929.

Simon-Vicente, F., Mas-Coma, S., Lopez-Roman, R., Tenora, F., Gallego, J., 1985. Biology of *Notocotylus neyrai* Gonzalez Castro, 1945 (Trematoda). Folia Parasitol. 32, 101—111.

Singh, K.S., 1957. On a new amphistome cercaria, C. lewerti, from India. Trans. Am. Microsc. Soc. 76, 366—370.

Singh, K.S., Lewert, R.M., 1959. Observations on the formation and chemical nature of metacercarial cysts of *Notocotylus urbanensis*. J. Infect. Dis. 104, 138—141.

Sinitsin, D.F., 1914. Neue tatsachen über die biologie der *Fasciola hepatica* L. Zentralblatt Bakteriol. Parasitenkd. Sec. 1 74, 280—285 (English translation in: Kean, B.H., Mott, K.E., Russell, A.J. (Eds.), Tropical Medicine & Parasitology: Classic Investigations, Cornell University Press, Ithaca, 1978, pp. 581—583.).

Smith, G., 1981. A three-year study of *Lymnaea truncatula* habitats, disease foci of Fascioliasis. Brit. Vet. J. 137, 398—410.

Smith, S.J., Hickman, J.L., 1983. Two new Notocotylid trematodes from birds in Tasmania and their life histories. Pap. Proc. Roy. Soc. Tasman. 117, 85—103.

Soliman, M.F.M., 2009. *Fasciola gigantica*: cercarial shedding pattern from *Lymnaea natalensis* after long-term exposure to cadmium at different temperatures. Exp. Parasitol. 121, 307—311.

Sonsino, P., 1892. Studi sui parassiti molluschi di aqua dolce dintorni di Cairo in Egitto. In: Festschrift zum Siebenzigsten Geburtstages Rudolf Leuckart's; dem verehrten Jubilar dargebracht von seinen dankbaren Schülern. Wilhelm Englemann, Leipzig, pp. 134—147.

Sprent, J.F.A., 1963. Parasitism: An Introduction to Parasitology and Immunology for Students of Biology, Veterinary Science, and Medicine. Bailliere, Tindall & Cox, London.

Steinberg, P.D., de Nys, R., 2002. Chemical mediation of colonization of seaweed surfaces. J. Phycol. 38, 621—629.

Stevens, T., 1996. The Importance of Spatial Heterogeneity in Organisms with Complex Life Cycles: Analysis of Digenetic Trematodes in a Salt Marsh Community (Ph.D. thesis). University of California, USA.

Stirewalt, M.A., 1971. Penetration stimuli for schistosome cercariae. In: Cheng, T.C. (Ed.), Aspects of the Biology of Symbiosis. University Park Press, Baltimore, pp. 1–23.

Stunkard, H.W., 1960. Studies on the morphology and life-history of *Notocotylus minutus* n.sp., a digenetic trematode from ducks. J. Parasitol. 46, 803–809.

Stunkard, H.W., 1966. The morphology and life-history of *Notocotylus atlanticus* n. sp., a digenetic trematode of eider ducks, *Somateria mollissima*, and the designation, *Notocotylus duboisi* nom. nov., for *Notocotylus imbricatus* (Looss, 1893) Szidat, 1935. Biol. Bull. 131, 501–515.

Stunkard, H.W., 1967a. The morphology, life-history, and systematic relations of the digenetic trematode *Uniserialis breviserialis* sp. nov., (Notocotylidae), a parasite of the bursa fabricius of birds. Biol. Bull. 132, 266–276.

Stunkard, H.W., 1967b. Studies on the trematode genus *Paramonostomum* Luhe, 1909 (Digenea: Notocotylidae). Biol. Bull. 132, 133–145.

Stunkard, H.W., 1981a. The life history, developmental stages, and taxonomic relations of the digenetic trematode *Lasiotocus minutus* (Manter, 1931) Thomas, 1959. Biol. Bull. 160, 146–154.

Stunkard, H.W., 1981b. The morphology, life history, and systematic relations of *Lasiotocus elongates* (Manter, 1931) Thomas, 1959 (Trematoda: Digenea). Biol. Bull. 160, 155–160.

Stunkard, H.W., Shaw, C.R., 1931. The effects of dilution of sea water on the activity and longevity of certain marine cercariae, with descriptions of two new species. Biol. Bull. 61, 242–271.

Stunkard, H.W., Cable, R.M., 1932. The life history of *Parorchis avitus* (Linton), a trematode from the cloaca of the gull. Biol. Bull. 62, 328–338.

Sudarikov, V.E., Karmanov, V.Y., 1980. The ability of molluscs to eliminate adolescariae (experimental study). Trudy Gel'mintol. Lab. Gel'minty Vod. Nazemn. Biotsen 30, 94–100 (In Russian).

Suhardono, Roberts, J.A., Copeman, D.B., 2006a. Distribution of metacercariae of *Fasciola gigantica* on rice straw. Trop. Anim. Health Prod. 38, 117–119.

Suhardono, Roberts, J.A., Copeman, D.B., 2006b. The effect of temperature and humidity on longevity of metacercariae of *Fasciola gigantica*. Trop. Anim. Health Prod. 38, 371–377.

Sukhdeo, M.V.K., Mettrick, D.F., 1986. The behavior of juvenile *Fasciola hepatica*. J. Parasitol. 72, 492–497.

Sulkin, S.D., 1986. Application of laboratory studies of larval behavior to fisheries problems. Can. J. Fish. Aquat. Sci. 43, 2184–2188.

Sulkin, S.D., 1990. Larval orientation mechanisms: the power of controlled experiments. Ophelia 32, 49–62.

Taylor, E.L., 1964. Fascioliasis and the Liver Fluke. Food and Agriculture Organization of the United States, Rome.

Taylor, E.L., Parfitt, J.W., 1957. Mouse test for the infectivity of metacercariae with particular reference to metacercariae in snail faeces. Trans. Am. Microsc. Soc. 76, 327–328.

Thieltges, D.W., Buschbaum, C., 2007. Vicious circle in the intertidal: facilitation between barnacle epibionts, a shell boring polychaete and trematode parasites in the periwinkle *Littorina littorea*. J. Exp. Mar. Biol. Ecol. 340, 90–95.

Thieltges, D.W., De Montaudouin, X., Fredensborg, B., Jensen, K.T., Koprivnikar, J., Poulin, R., 2008. Production of marine trematode cercariae: a potentially overlooked path of energy flow in benthic systems. Mar. Ecol. Prog. Ser. 372, 147–155.

Thommen, G.H., Westlake, D.F., 1981. Factors affecting the distribution of populations of *Apium nodiflorum* and *Nasturtium officinale* in small chalk streams. Aquat. Bot. 11, 21–36.

Threlkeld, S.T., Cchiavelli, D.A., Willey, R.L., 1993. The organization of zooplankton epibiont communities. Trends Ecol. Evol. 8, 317–321.

Tielens, A.G.M., Van den Heuvel, J., Van den Bergh, S.G., 1982. Changes in energy metabolism of the juvenile *Fasciola hepatica* during its development in the liver parenchyma. Mol. Biochem. Parasitol. 6, 277–286.

Tolstenkov, O.O., Akimova, L.N., Terenina, N.B., Gustafsson, M.K.S., 2012. The neuromuscular system in continuously swimming cercariae from Belarus. II. Echinostomata, Gymnocephala and Amphistomata. Parasitol. Res. 111, 2301–2309.

Torgerson, P., Claxton, J., 1999. Epidemiology and control. In: Dalton, J.P. (Ed.), Fasciolosis. CABI Publishing, Wallingford, pp. 113–149.

Tripathi, A.P., Singh, V.K., Sing, D.K., 2014. Response of *Fasciola gigantica* cercaria exposed to different photo and chemo stimulants. J. Biol. Earth Sci. 4, B36–B42.

Ueno, H., Yoshihara, S., 1974. Vertical distribution of *Fasciola gigantica* metacercariae on stems of rice plant grown in a water pot. Natl. Inst. Anim. Health Q. (Tokyo) 14, 54–60.

Valero, M.A., Mas-Coma, S., 2000. Comparative infectivity of *Fasciola hepatica* metacercariae from isolates of the main and secondary reservoir animal host species in the Bolivian Altiplano high human endemic region. Folia Parasitol. 47, 17–22.

Vareille-Morel, C., Rondelaud, D., 1991. Les metacercaires flottantes de *Fasciola hepatica* L. etude experimentale de facteurs sur leur formation. Bull. Soc. Franc. Parasitol. 9, 81–85.

Vareille-Morel, C., Esclaire, F., Hourdin, P., Rondelaud, D., 1993a. Internal metacercarial cysts of *Fasciola hepatica* in the pulmonate snail *Lymnaea truncatula*. Parasitol. Res. 79, 259–260.

Vareille-Morel, C., Dreyfuss, G., Rondelaud, D., 1993b. Premières données sur la dispersion et le devenir des métacercaires flottantes de *Fasciola hepatica* L. Bull. Soc. Franc. Parasitol. 11, 63–69.

Vareille-Morel, C., Dreyfuss, G., Rondelaud, D., 1994a. *Fasciola gigantica* Cobbold et *F. hepatica* Linne: les variations numeriques des kystes flottants en function de l'espece de la limnee et de sa taille lors de l'exposition aux miracidiums. Bull. Soc. Franc. Parasitol. 12, 161–166.

Vareille-Morel, C., Dreyfuss, G., Rondelaud, D., 1994b. *Fasciola hepatica* Linne: relations entre l'ordre de sortie des cecaires a partir de *Lymnaea truncatula* Muller et la formation des kystes flottants ou fixes. Bull. Soc. Franc. Parasitol. 12, 55–60.

Varma, A.K., 1961. Observations on the biology and pathogenicity of *Cotylophoron cotylophorum* (Fischoeder, 1901). J. Helminthol. 35, 161–168.

Varma, T.K., Prasad, A., 1998. Chromoaffinity of *Paramphistomum epiclitum* (Fischoeder, 1904) cercariae. Riv. Parassitol. 59, 199–204.

Velasquez, C.C., 1969. Life history of *Paramonostomum philippinensis* sp. n. (Trematoda: Digenea: Notocotylidae). J. Parasitol. 55, 289–292.

Vernberg, W.B., Vernberg, F.J., 1971. Respiratory metabolism of a trematode metacercaria and its host. In: Cheng, T.C. (Ed.), Aspects of the Biology of Symbiosis. University Park Press, Baltimore, pp. 91–102.

Verma, A., Singh, S.N., 2006. Biochemical and ultrastructural changes in plant foliage exposed to auto-pollution. Environ. Monit. Assess. 120, 585–602.

Vieira, L.Q., Oliveira, M.R., Neumann, E., Nicoli, J.R., Vieira, E.C., 1998. Parasitic infections in germfree animals. Braz. J. Med. Biol. Res. 31, 105–110.

Vignoles, P., Alarion, N., Bellet, V., Dreyfuss, G., Rondelaud, D., 2006. A 6- to 8-day periodicity in cercarial shedding occurred in some *Galba truncatula* experimentally infected with *Fasciola hepatica*. Parasitol. Res. 98, 385–388.

Volaire, F., Lelievre, F., 2001. Drought survival in *Dactylis glomerata* and *Festuca arundinacea* under similar rooting conditions in tubes. Plant Soil 229, 225–234.

Wahl, M., 1989. Marine epibiosis. I. Fouling and antifouling: some basic concepts. Mar. Ecol. Prog. Ser. 58, 175–189.

Wahl, M., Kroge, K., Lenz, M., 1998. Non-toxic protection against epibiosis. Biofouling 12, 205–226.

Wahl, M., Hay, M.E., 1995. Associated resistance and shared doom: effects of epibiosis on herbivory. Oecologia 102, 329–340.

Wardle, W.J., 1988. A Bucephalid larva, *Cercaria pleuromerae* n. sp. (Trematoda: Digenea), parasitizing a deepwater bivalve from the Gulf of Mexico. J. Parasitol. 74, 692−694.
Warner, G.F., 1977. The Biology of Crabs. Elek, London.
Warren, K.S., Peters, P.A., 1968. Cercariae of *Schistosoma mansoni* and plants: attempts to penetrate *Phaseolus vulgaris* and *Hedychium coronarium* produces a cercaricide. Nature 217, 647−648.
Weis, J.S., Cristini, A., Rao, K.R., 1992. Effects of pollutants on molting and regeneration in Crustacea. Am. Zool. 32, 495−500.
Weng, Y.L., Zhuang, Z.L., Jiang, H.P., Lin, G.R., Lin, J.J., 1989. Studies on ecology of *Fasciolopsis buski* and control strategy of fasciolopsiasis. Chin. J. Parasitol. Parasit. Dis. 7, 108−111 (In Chinese).
Wescott, R.B., 1970. Metazoa-Protozoa-Bacteria interrelationships. Am. J. Clin. Nutr. 23, 1502−1507.
Wesenberg-Lund, C., 1934. Contributions to the development of the Trematoda Digenea. Part II. The biology of the freshwater cercariae in Danish freshwaters. Mem. Acad. Roy. Sci. Lett. Dan. Cph. 5 (3), 1−223.
West, A.F., 1961. Studies on the biology of *Philopthalmus gralli* Mathis and Leger, 1910 (Trematoda: Digenea). Am. Midl. Nat. 66, 363−383.
Wetzel, E.J., Shreve, E.W., 2003. The influence of habitat on the distribution and abundance of metacercariae of *Macravestibulum obtusicaudum* (Pronocephalidae) in a small Indiana stream. J. Parasitol. 89, 1088−1090.
Whittington, I.D., Cribb, B.W., 2001. Adhesive secretions in the Platyhelminthes. Adv. Parasitol. 48, 101−224.
Willey, R.L., Cantrell, P.A., Threlkeld, S.T., 1990. Epibiotic euglenoid flagellates increase the susceptibility of some zooplankton to fish predation. Limnol. Oceanogr. 35, 952−959.
Willey, R.L., Willey, R.B., Threlkeld, S.T., 1993. Planktivore effects on zooplankton epibiont communities: epibiont pigmentation effects. Limnol. Oceanogr. 38, 1818−1822.
Williams, M.O., 1969. Cercaria *Parorchis acanthi* in marine molluscs from Sierra Leone with notes on survival of the cercaria and development of the metacercaria in the definitive host. Rev. Zool. Bot. Afr. 79, 265−272.
Wings, O., 2007. A review of gastrolith function with implications for fossil vertebrates and a revised classification. Acta Palaeontol. Pol. 52, 1−16.
Wisniewski, L.W., 1937. Entwick-lungszyklus und Biologie von *Parafasciolopsis fasciolaemorpha* Ejsm. Mem. Acad. Pol. Sci. Lett. Ser. B 11, 1−113.
Woodin, S.A., 1991. Recruitment of infauna: positive or negative cues? Am. Zool. 31, 797−807.
Wunder, W., 1932. Untersuchungen über pigmentierung und encystierung von cercarien. Z. Morph. Ökol. Tiere 25, 336−352.
Xu, Z., Burns, C.W., 1991. Effects of the epizoic ciliate, *Epistylis daphnia*, on growth, reproduction, and mortality of *Boeckella triarticulata* (Thomson) (Copepoda: Calanoida). Hydrobiologia 209, 183−189.
Yadav, S.C., Gupta, S.C., 1988. On the viability of *Fasciola gigantica* metacercariae ingested by *Lymnaea auricularia*. J. Helminthol. 62, 303−304.
Yamaguti, S., 1975. A Synoptical Review of Life-Histories of Digenetic Trematodes of Vertebrates. Keigaku Publishing Co, Tokyo.
Yokogawa, M., 1964. *Paragonimus* and paragonimiasis. Prog. Med. Parasitol. Jap. 1, 63−218.
Yoshihara, S., Ueno, H., 2004. Ingestion of *Fasciola gigantica* metacercariae by the intermediate host snail, *Lymnaea ollula*, and infectivity of discharged metacercariae. Southeast Asian J. Trop. Med. Public Health 35, 535−539.

Young, C.M., 1995. Behavior and locomotion during the dispersal phase of larval life. In: McEdward, L. (Ed.), Ecology of Marine Invertebrate Larvae. CRC Press, Boca Raton, pp. 249–277.

Zimmer-Faust, R.K., Tamburri, M.N., 1994. Chemical identity and ecological implications of a waterborne, larval settlement cue. Limnol. Oceanogr. 39, 1075–1087.

Zimmer, R.K., Fingerut, J.T., Zimmer, C.A., 2009. Dispersal pathways, seed rains, and the dynamics of larval behavior. Ecology 90, 1933–1947.

CHAPTER TWO

Cross-Border Malaria: A Major Obstacle for Malaria Elimination

Kinley Wangdi[*,§,1], Michelle L. Gatton[¶], Gerard C. Kelly[*], Archie CA. Clements[*]

[*]The Australian National University, Research School of Population Health, College of Medicine, Biology and Environment, Canberra, ACT, Australia
[§]Phuentsholing General Hospital, Phuentsholing, Bhutan
[¶]Queensland University of Technology, School of Public Health & Social Work, Brisbane, Qld, Australia
[1]Corresponding author: E-mail: dockinley@gmail.com

Contents

1. Introduction	80
2. Patterns of Movement	82
2.1 Migration for work opportunities	86
2.2 Visiting friends and relatives	87
2.3 Treatment-seeking behaviour in border areas	88
2.4 Displacement due to conflict and major development projects	89
3. Epidemiological Drivers of Malaria in Border Areas	89
3.1 Misalignment of programmatic approaches	90
3.2 Forests and deforestation	90
3.3 Socioeconomic factors	91
4. Way Forward	92
4.1 International collaboration	92
4.2 Surveillance-response and cross-border initiatives	94
4.3 Strengthening of preventive measures for cross-border malaria	96
4.4 Technological solutions to support operational decision making and surveillance response	96
5. Conclusion	97
Contributors	98
References	99

Abstract

Movement of malaria across international borders poses a major obstacle to achieving malaria elimination in the 34 countries that have committed to this goal. In border areas, malaria prevalence is often higher than in other areas due to lower access to health services, treatment-seeking behaviour of marginalized populations that typically inhabit border areas, difficulties in deploying prevention programmes to hard-to-reach communities, often in difficult terrain, and constant movement of people across porous national boundaries. Malaria elimination in border areas will be challenging and key to addressing the challenges is strengthening of surveillance activities for rapid identification of any importation or reintroduction of malaria. This could involve taking advantage of technological advances, such as spatial decision support systems, which can be deployed to assist programme managers to carry out preventive and reactive measures, and mobile phone technology, which can be used to capture the movement of people in the border areas and likely sources of malaria importation. Additionally, joint collaboration in the prevention and control of cross-border malaria by neighbouring countries, and reinforcement of early diagnosis and prompt treatment are ways forward in addressing the problem of cross-border malaria.

1. INTRODUCTION

Globally, an estimated 3.4 billion people were at risk of malaria in 2012, with populations living in sub-Saharan Africa having the highest risk of acquiring malaria (World Health Organization, 2013). Approximately 80% of cases and 90% of deaths are estimated to occur in the World Health Organization (WHO) African Region, with children under five years of age and pregnant women being most severely affected (World Health Organization, 2012; Casalino et al., 2002; Martens and Hall, 2000; World Health Organization, 2013). WHO estimated that 207 million cases of malaria occurred in 2012 (uncertainty range 135–287 million) and 627,000 deaths (uncertainty range 473,000–789,000) (World Health Organization, 2013). Deaths attributed to malaria have declined by 32% between 2004 and 2010 (Murray et al., 2012). This reduction has most likely been a result of the combined effects of economic development in endemic countries, urbanization and unprecedented financial support for malaria interventions from donors and the associated scaling up of malaria interventions. In sub-Saharan Africa, there was a 66-fold increase in the amount of official development assistance disbursed for malaria control, from $9.8 million in 2002 to $651.7 million in 2008 (Akachi and Atun, 2011). Major funders include Roll Back Malaria, the Global Fund to Fight AIDS, Tuberculosis and Malaria, the US President's Malaria Initiative and the World Bank's International Development Association (Feachem et al., 2010a) and funding from the Bill and Melinda Gates Foundation has been transformational in driving malaria elimination research. The increased funding has supported scaling up of preventive activities such as provision of long-lasting insecticide-treated bed nets (LLINs) and indoor residual spraying (IRS) as the principal vector control measure, as well as improving timely diagnosis using rapid diagnostic tests (RDTs) and providing effective treatment with artemisinin-based combination therapy (Gueye et al., 2012; Anderson et al., 2011). As a result of these gains, and renewed global interest, 32 of the 99 malaria-endemic countries are now pursuing an elimination strategy, with the remaining 67 aiming to control malaria (Das and Horton, 2010; Feachem et al., 2010a, 2010b).

The second-generation Global Malaria Action Plan (GMAP2) for the period 2016—2025 has now commenced. The GMAP2 aims to accelerate progress in malaria elimination at global, regional and country levels and serve as a major advocacy instrument for the achievement of a malaria-free world. A three-part strategy to eliminate malaria has been developed and is now widely endorsed: (1) aggressive control in highly endemic countries, to lower transmission and mortality in countries that have the highest burden of disease and death; (2) progressive elimination of malaria from the endemic margins, to 'shrink the malaria map' and (3) research into vaccines and improved drugs, diagnostics, insecticides and other interventions, and into delivery methods that reach all at-risk populations (Feachem et al., 2010a; Roll Back Malaria, 2008; Breman and Brandling-Bennett, 2011; Feachem and Sabot, 2008; Mendis et al., 2009). The defining aspects of malaria elimination are outlined in Panel 1.

Although great gains have been made in reducing the overall burden of malaria, impact from elimination and control efforts proves more difficult in areas near international borders. The specific environmental (including physical, social and geopolitical), anthropological, administrative and geographic characteristics of border areas impact uniquely on the epidemiology and control of malaria, resulting in coinage of the terms 'border malaria' and 'cross-border malaria'. Here, we apply the term cross-border malaria to encompass malaria transmission as a result of cross-border movement of people or vectors, in addition to the epidemiological situation that occurs in relation to malaria in areas adjacent or near to international borders (i.e. border malaria).

Cross-border malaria is difficult to manage due to political, economic and geographic constraints (Xu and Liu, 2012). Factors such as frequent movement of humans and vectors across borders, lack of responsibility of individual countries in the border endemic areas and relatively poor access to health care and preventative measures, particularly for mobile populations, leave space for reservoirs of infection that can lead to continued transmission of malaria and vulnerability to malaria outbreaks and epidemics (Gueye et al., 2012).

> **Panel 1 Malaria elimination defining activities**
>
> Malaria elimination requires:
> - evidence-based data on the achievement of successful malaria control;
> - sufficient evidence that transmission can be interrupted by scaling up planned interventions;
> - clearly defined responsibilities for management, including decentralized authority and enforcement of regulatory and disciplinary measures;
> - effective systems to ensure coordination between public, private and community-based agencies and services, and to implement cross-border programmes;
> - intensive joint inter-sectoral efforts;
> - adequate pre- and in-service training of service providers and high-quality supervision/mentoring;
> - sustained advocacy, social mobilization, health education and behavioural change communication to support the preparation and implementation of the elimination programme;
> - the existence of a monitoring, evaluation and surveillance plan able to timely measure progress, including assessments by independent team(s);
> - long-term predictable and sustainable funding available to support planned and unexpected expenses;
> - eventually, systems in place for effective vigilance to prevent reintroduction.
>
> *WHO. 2007. Malaria elimination- a field manual for low and moderate endemic countries. World Health Organization, Geneva.*

The aim of this review is to present a compilation of evidence in the available literature on the impact of cross-border malaria on elimination efforts. Drivers of cross-border malaria are described and measures to prevent or mitigate cross-border malaria are discussed. The review for this paper was carried out using the search engines PubMed, Medline and Google Scholar. The key search words were malaria, cross-border, migration, international borders and malaria elimination. We reviewed all relevant articles written in English.

2. PATTERNS OF MOVEMENT

Cross-border malaria encompasses malaria transmission along international borders as a result of interconnections between human settlements and population movement, including localized border crossings and population migration over larger distances (Guerra et al., 2006; Olson et al., 2010).

> **Panel 2 Different types of movement across borders**
> - Circulation encompasses a variety of movements involving no longstanding change in residence.
> - Migration involves a permanent change of residence.
> - Daily circulation involves leaving the place of residence for up to 24 h.
> - Periodic circulation may vary from one night to one year, but is usually shorter than for seasonal circulation.
> - Seasonal circulation involves a period in which persons or groups are absent from their permanent homes during a season or seasons of the year.
> - Long-term circulation involves an absence from the home for longer than one year.
> - Active transmitters 'source' harbour the parasite and transmit the disease when they move to new areas known as 'sinkers', which may have a low-level or sporadic transmission.
> - Passive acquirers are exposed to the disease through the movement from one environment to another; they may have a low level of immunity, which increases their risk of clinical malaria.

Border crossings can be defined as movements of local people between countries that occur with or without passing border control checkpoints. Cross-border migration can be defined as the movement of people from a country of origin to a destination country with or without passing border control checkpoints for either short-term or long-term immigration with different channels of migration (Panels 2—5, Koyadun and Bhumiratana, 2005; Bhumiratana et al., 2010).

Population movements can be differentiated by their temporal and spatial dimensions. Temporal dimensions include circulation and migration. Circulation encompasses a variety of movement, usually short-term and cyclical and involving no longstanding change in residence. Migration

> **Panel 3 Different approaches in tackling cross-border malaria**
> Joint colourations targeting malaria control and prevention between the countries that share the border.
> Robust surveillance system for identifying the importation of malaria across borders and reintroduction of malaria after successful elimination.
> Administration of antimalarial drugs with the use of protective measures.
> SDSS could be used to target and coordinate cross-border malaria interventions.
> Use of mobile technology in assessing the movement of people across borders.

Panel 4 Summary of different interventions to address cross-border malaria

Approaches in tackling cross-border malaria	Advantages	Limitations
Joint collaboration	Prompt sharing of cross-border data. Tackling any possibilities of out breaks.	Requires time to build trust among the health workers of the different countries.
Administration of antimalarial drugs with the use of protective measures for migrants	Avert the risk of spreading and introducing malaria into the naive population. Radical cure of malaria. Prevent development of drug-resistant malaria. The chemoprophylaxis can prevent malaria transmission from sources to sinks.	Ineffective in the mobile population, which involves in crossing border frequently daily. The cost of diagnosis and treatment will be the main barrier if treatment is not provided free.
Surveillance systems	The robust surveillance at points of entry of higher transmission will facilitate swift treatment and follow-up of infected individuals. Once the interruption of transmission has been achieved, surveillance systems will play an important role in the prevention of reintroduction.	The differences in the surveillance system among the countries need to be resolved so that neighbouring countries operate a similar interface. Not easy to identify points of entry in remote border areas.
Spatial decision support system (SDSS)	Conduct high-resolution surveillance and able to locate and classify active transmission foci. A Regional SDSS framework could provide malaria data and malaria transmission across borders. This information could be used by the relevant partners to target and coordinate cross-border malaria interventions.	The technical knowhow would be the main barrier while implementing SDSS. SDSS may not be able to work well among transient/floating population.
Mobile telecommunication on tracking cross-border malaria	Quantify the volume of the people crossing the border areas. Movement patterns derived from phone records can inform on the likely sources and rates of malaria importation.	Use of mobile technology in tracking cross-border malaria is a new concept. Restricted to areas with mobile network coverage; Access to proprietary data will probably be difficult; data quality and completeness potentially low.

> **Panel 5 Search strategy and selection criteria**
> The review for this paper was carried out using online search engines including PubMed, Medline and Google Scholar. The key search words were 'malaria', 'cross-border', 'migration of people across international borders' and 'malaria elimination'. We reviewed all the articles written in English with preference for recent publications.

movements involve a permanent change of residence (Prothero, 1977; Stoddard et al., 2009). Circulatory movement can be subdivided into daily, periodic, seasonal and long term. Daily circulation involves leaving a place of residence for up to 24 h. Periodic circulation may vary from one night to one year but is usually of a shorter duration than seasonal circulation. Seasonal circulation involves a period in which persons or groups are absent from their permanent homes during one or more seasons of the year. With regard to long-term circulation, there is absence from the home for longer than one year, but with maintenance of close social and economic ties with the home area (Wolpert, 1965; Martens and Hall, 2000; Roseman, 1971; Stoddard et al., 2009; Prothero, 1977; Pindolia et al., 2012).

People cross international borders for a number of reasons, including migration for work opportunities, visiting friends and relatives (VFRs), tourism, travel for business purposes or cross-border trade, social relations, cultural exchanges (pilgrimages, festivities, fairs, etc.) and displacement as a result of natural and artificial calamities (e.g. wars) and major development projects, such as construction of dams. Some of these movements increase exposure to malaria parasites, particularly in forests or areas of deforestation, where occupational exposure may occur.

It is difficult to obtain basic data on key variables, such as the actual numbers of movements of people across borders, or for such data to be broken down by movement type (e.g. border crossings versus cross-border migrations) (Khamsiriwatchara et al., 2011). For example, migrations across the international borders of Yunnan Province, China, which shares >4000 km of border with Myanmar, Lao People's Democratic Republic (PDR) and Vietnam, take place unchecked (Hu et al., 1998; Clements et al., 2009). Similarly, unmonitored migration of people across the border from Myanmar into Bangladesh jeopardizes the control efforts in Bangladesh (Reid et al., 2010) and imported infections from Yemen into Saudi Arabia continue to challenge Saudi elimination efforts (Alkhalife, 2003).

Movement of people across international borders has contributed to maintaining high transmission at hotspots adjacent to border points (Clements et al., 2009; Carme, 2005). A major challenge to sustaining elimination is addressing the potential reintroduction of cases, either via border areas or from migrant populations (Tatem and Smith, 2010). Nearly 20% of malaria cases treated in Iran in 2006 originated in Pakistan (Reza et al., 2009). Local transmission of malaria in the United Arab Emirates (UAE) came to an end in 1997, and no autochthonous cases were reported from 1998 to 2004. Therefore, the UAE was certified as a malaria-free country. However, there was importation of malaria into the UAE from the neighbouring countries (Sultan et al., 2009).

2.1 Migration for work opportunities

The majority of migrants cross borders in search of better economic, work and social opportunities. Economic migrants are the world's fastest growing group of migrants. Economic motivations are the main reasons for people to migrate from countries with high levels of malaria to malaria-eliminating countries, impeding malaria elimination efforts in those countries (Carme, 2005; Davin and Majidi, 2009; Kitvatanachai et al., 2003; Wangdi et al., 2011). Economic migration is exacerbated when there are substantial differences in the economic development and job opportunities in neighbouring countries. For instance, economic stagnation in Myanmar and rapid economic development in Thailand have stimulated migration from Myanmar to Thailand (Carrara et al., 2013; Delacollette et al., 2009; Huguet and Punpuing, 2005; Khamsiriwatchara et al., 2011; Wangroongsarb et al., 2012), while temporary migration of seasonal workers from Cambodia to Thailand seems to be a key factor responsible for the malaria problem along the Cambodian—Thailand border (Hoyer et al., 2012; Kitvatanachai et al., 2003). It is estimated that 50—70% of all reported malaria cases in Argentina are linked to migration, in particular movement across the border from Bolivia; this migration is fuelled by economic growth on the Argentine side and is not well controlled due to a porous border between the two countries (Gueye et al., 2012). Malaria increased substantially in French Guiana due to the influx of Brazilians to work in gold mining (Carme, 2005). Economic motivations are the main reasons for Afghans to migrate to Pakistan. As high as 64.6% of Afghan migrants crossing into Pakistan cited lack of work in Afghanistan as the main factor leading them to Pakistan (Davin and Majidi, 2009). Economic migration also happens beyond countries sharing common international borders. For instance, imported malaria in Jiangsu Province, China, from 2001 to 2011 accounted for up to 12.4% of cases, mainly imported by Chinese nationals from African countries as a result of economic migration (Liu et al., 2014).

The resurgence of malaria in Swaziland in early 1970 occurred as a result of the migration of sugar cane workers from malaria-endemic Mozambique (Martens and Hall, 2000; Packard, 1986). More recently, the current migration of labourers into Swaziland from Mozambique is likely to be a challenge for Swaziland's stated plan of malaria elimination by 2015 (Koita et al., 2013). The rapid rise in malaria incidence in Brazil in the late 1970s and early 1980s was attributed to the influx of malaria-infected migrants from endemic Bolivia (Cruz Marques, 1987). The resurgence of malaria in Costa Rica resulted due to the development of the banana industry in which workers were moved from endemic areas into areas with increased suitability for vector breeding (Najera et al., 1998). The oil-exporting countries of the Middle East have attracted a large number of semiskilled workers from malarious countries such as India, Pakistan and Indonesia, who are a source of malaria introduction (Schultz, 1989). The importation of malaria to Kuwait occurs mostly from the Indian subcontinent (Hira et al., 1988, 1985; Iqbal et al., 2003). Saudi Arabia is an attractive employer of skilled workers from malaria-endemic countries such as Iran, Pakistan and India, as well as east Africa (Bruce et al., 2000; Babiker et al., 1998). The main source of malaria cases in the UAE is from Pakistan and neighbouring Oman, including families of UAE nationals living across the border in Oman (Dar et al., 1993). These examples highlight the important role that economic migrations have in re-establishing malaria in areas where control efforts had previously been successful.

2.2 Visiting friends and relatives

Ethnic groups are often spread across borders, and people may cross the international border to meet relatives and friends (Pongvongsa et al., 2012; Noor et al., 2013). Immigrant VFRs frequently return to visit family members whom they had left behind or to introduce new additions to the family of origin. Last-minute travel to visit sick relatives or attend funerals is common, allowing little time for provision and receipt of pretravel advice on malaria prevention. Other travel reasons include finding a spouse, locating missing family, or returning for traditional or cultural ceremonies (Xu and Liu, 1997). Many VFRs stay in family settings in which they may encounter suboptimal housing conditions and increased malaria risk (Bacaner et al., 2004; Scolari et al., 2002; Muentener et al., 1999; Fulford and Keystone,

2005; Di Perri et al., 1994; Barnett et al., 2010; Fenner et al., 2007; Froude et al., 1992; Wagner et al., 2013). VFRs may encounter barriers such as lack of information on services, language, trust of health systems, concerns on their legal status and cost of malaria chemoprophylaxis, which may limit their access to travel clinics (Bacaner et al., 2004; Stager et al., 2009). Migrant VFRs may be exposed to risk of malaria as they visit their families in rural areas with higher malaria transmission rates (Schlagenhauf et al., 2003).

2.3 Treatment-seeking behaviour in border areas

The porous nature of many borders encourages people to migrate and seek treatment across borders. For example, malaria patients from the state of Assam, India, often travel to hospitals in neighbouring Bhutan to receive treatment because treatment is free on the Bhutanese side of the border (Yangzom et al., 2012). Due to poor health infrastructure in Nepal, a large number of people from the plains and hills in the south of the country travelled in the past to hospitals in India to access health care. However, in the last few decades, Nepal has been able to develop health facilities in the country, particularly in the plains, with the establishment of regional, zonal and district hospitals with modern medical facilities. This has resulted in the large-scale reverse flow of people from India seeking treatment in these hospitals (Kansakar, 2001).

Migrant workers are less likely than the general population to get blood tested for malaria parasites and get radical treatment (Hiwat et al., 2012). Migrant workers and border people have often demonstrated suboptimal health-seeking behaviours and often self-medicate. Malaria treatment in the border areas is often inadequate. Inadequate public health facilities in border areas lead local populations to seek treatment from private health professionals, many of whom provide counterfeit or substandard antimalarial drugs, or monotherapies, resulting in an increased risk of antimalarial drug resistance (Pongvongsa et al., 2012; Wijeyaratne et al., 2005). Thus, these groups are among the principal contributors to the emergence of multi-drug resistant, which is a particular problem along the Thailand—Myanmar and Thailand—Cambodia borders (Satitvipawee et al., 2012; Thimasarn, 2003; WHO, 2010). Gold miners in French Guiana do not seek malaria treatment in their country due to their illegal status and high local transportation costs; rather, they seek diagnosis and treatment in Suriname. Low accessibility to diagnosis and treatment for these gold miners has resulted in a flourishing black market for antimalarial drugs, often with insufficient quality (Hiwat et al., 2012).

2.4 Displacement due to conflict and major development projects

The World Bank estimates >1.5 billion people live in violent, conflict-affected countries (The World Bank, 2012). Movement of displaced people, including refugees, and soldiers as a result of conflict or war has been implicated as a cause of malaria resurgence in Bangladesh, Vietnam, Sri Lanka, Sudan and Azerbaijan. Decades of internal conflict in Myanmar have resulted in massive population displacement, and >150,000 refuges now live in camps in Thailand (Carrara et al., 2013). Similarly, the Islamic Republic of Iran hosts around 1.5–2 million Afghani refugees (Basseri et al., 2010). These displaced people play an important role in the transmission of malaria due to inadequate control and preventive measures. The displaced people face unreliable access to basic services including health care (Williams et al., 2013). People living in conflict zones, such as the Karen, have higher mortality rates irrespective of malaria incidence (Lee et al., 2006).

The construction of China's Three Gorges Dam resulted in the relocation of 1.3 million people. There has been an epidemic of locally transmitted malaria among residents at the dam site in 1996, and this could recur and spread (Jackson and Sleigh, 2000). The construction of the Bargi dam in India saw a 2.4-fold increase in malaria cases and a more than fourfold increase in annual parasite incidence among children in the villages closer to the dam compared with more distant villages. In addition, there was a strong increase in prevalence in the partially submerged villages (Singh et al., 1999; Singh and Mishra, 2000). Dam construction, irrigation and other development projects, urbanization and deforestation have all resulted in changes in vector population densities and emergence of new diseases and re-emerge old diseases (Walsh et al., 1993; Patz et al., 2000; Gratz, 1999; Keiser et al., 2005).

3. EPIDEMIOLOGICAL DRIVERS OF MALARIA IN BORDER AREAS

Malaria control in border areas is often more difficult than in central and non-border areas due to heavily forested, mountainous and inaccessible terrain, and because of unregulated population movements across the border (Xu and Liu, 2012). In addition, many border areas are inhabited by ethnic minorities (Prothero, 1999; Erhart et al., 2005) with limited formal education (Erhart et al., 2007) and less access to health education efforts. The impact of different national policies for control and prevention in neighbouring countries is potentially causing a lack of political will to invest in border areas.

3.1 Misalignment of programmatic approaches

Differences in programmatic approaches between neighbouring countries commonly occur making the coordination of control and preventive measures in the border areas challenging. One such example is the Laos–Vietnam border where malaria control on the Laos side is based on distribution of LLINs but on the Vietnamese side it relies mainly on IRS of insecticides (Anh et al., 2005; Hung le et al., 2002). There are also differences in malaria diagnosis and treatment between the two countries. RDTs are mostly used for diagnosis in Laos, while Vietnam uses microscopy as a rule.

Even where the approaches are similar between neighbouring countries, the specific drugs or chemicals used can influence their effectiveness due to parasite or vector resistance. For example, deltrametrin (a synthetic pyrethoid) is used for IRS in Bhutan, whereas dichlorodiphenyltrichloroethane (DDT) is still used in the neighbouring states of Assam, India, even though there are reports of vector resistance to DDT (Dev et al., 2006; Wangdi et al., 2010; Mittal et al., 2004). Effective control or elimination requires both countries across the international boundary to be committed to malaria interventions. In addition, control and preventive activities including IRS need to be synchronized to achieve maximum benefit.

3.2 Forests and deforestation

Both the presence of forests and occurrence of deforestation have impact on increasing malaria risks and transmission in border areas. Populations in border areas are at a greater risk of malaria infections because they frequently visit forestlands, forest fringe areas or forested plantations at or near the border (Chaveepojnkamjorn and Pichainarong, 2004). Forest-related activities and factors related to poverty are major drivers of malaria incidence in Viet Nam (Manh et al., 2011; Erhart et al., 2004). Many species of *Anopheles* mosquitoes that transmit malaria are common fauna of natural forests and forested plantations in border areas. Border populations are particularly at a risk of occupational exposure to malaria through working in crop plantations, forestry, mining, development projects and tourism (Pichainarong and Chaveepojnkamjorn, 2004). Occupational exposures affect the age profile of malaria infections, for example, in forest fringe villages, adult rather than childhood infections are more prevalent due to forest-related activities of workmen, such as logging, bamboo cutting, charcoaling, foraging and overnight stays in the forests (Dysoley et al., 2008). Migration of the population working in the forest and forest fringe can result in spread via carriers to new areas previously not known for malaria transmission (Wisit Chaveepojnkamjorn, 2005). These result in an increase in human infection, not only within the mobile population but also within the fixed population, to which the migrants return periodically.

Changes in land cover associated with economic activities can enhance contact with mosquitoes and thereby increase malaria transmission. Deforestation has occurred in many malaria-endemic areas as a result of colonization and settlement programmes, logging, increased large-scale agricultural activities, mining, the building of hydropower schemes and the collection of wood for fuel. Deforestation activities lead to a host of influences on the distribution and prevalence of vector-borne diseases. New habitats for *Anopheles darlingi* mosquitoes are created through the formation of large ponds and presence of leaf litter, algae and emergent grasses due to deforestation or activities associated with it. This has led to malaria epidemics in South America (Olson et al., 2010; Vittor et al., 2006, 2009). Increased deforestation in Brazil leads to increased malaria cases in Mancio Lima County (Olson et al., 2010). Populations residing within or near the fragmented forests are at a higher risk of malaria because of increased contact with the vectors at the forest edges and reduced biodiversity. Continued deforestation throughout the world will likely continue to result in increased vector-borne diseases (Guerra et al., 2006).

3.3 Socioeconomic factors

Residual transmission in some malaria-eliminating countries is concentrated in a few hard-to-reach populations, of which mobile populations within border areas are included. These populations often have unofficial status and few economic resources, and can be difficult to locate for the purposes of control and effective treatment of malaria (Stern, 1998).

Ethnic minorities in border areas often have limited formal education, impeding health promotion efforts, resulting in prevalent risk behaviours such as improper use of insecticide-treated nets and other protective measures, and limiting access to healthcare (Prothero, 1999; Erhart et al., 2005). Such groups are typically impoverished and mobile, often driven to more remote areas by marginalization and safety concerns (Martens and Hall, 2000; Chuquiyauri et al., 2012; Prothero, 1995; Xu and Liu, 1997). They might avoid accessing the health systems because of fear of unwanted attention from government authorities, thus making monitoring and treatment of their malaria difficult (Hiwat et al., 2012). Distinctive ethnic minority groups can vary in terms of cultural practices, languages and life styles that are of relevance to malaria risk, including the practice of staying in the forest overnight.

Poverty serves as a motivating reason for people to seek income from occupational activities associated with forests and mining that might expose them to higher risks (Chaveepojnkamjorn and Pichainarong, 2004). Such activities may be illegal, and as a result, their members often face substantial barriers to healthcare access (Hiwat et al., 2012). For the poor, living conditions are associated with inadequate housing and overcrowding, which increase the risk of malaria. Houses are hastily constructed and are often made of locally available materials. Inadequate housing might allow mosquitoes to enter more easily than well-constructed housing with screened windows. The risk of getting malaria has been shown to be greater for inhabitants of the poorest type of house construction (incomplete, mud, or palm walls and palm thatched roofs) compared to houses with complete brick and plaster walls and tiled roofs (Gamage-Mendis et al., 1991; Konradsen et al., 2003; Lindsay et al., 2002).

4. WAY FORWARD
4.1 International collaboration

Malaria control strategies and policies as well as the quality and management of the health care systems and conventions in data collection may differ across national borders, making cross-border collaboration difficult (Pongvongsa et al., 2012). However, the phenomenon of cross-border malaria provides a strong rationale to develop harmonized cross-border programmes in conjunction with national efforts (Delacollette et al., 2009). The philosophy of cross-border or regional collaboration has been well adopted in different regions, and the results have been positive. One example is the Lubombo Spatial Development Initiative (LSDI) between South Africa, Swaziland and Mozambique. The LSDI was made possible as a result of political commitment through the signing of a protocol of understanding by the head of three states, which created a platform for regional cooperation and delivery. The malaria control programme of the LSDI aimed to achieve maximum effectiveness of malaria control in the highest-risk areas of South Africa and Swaziland bordering Mozambique. These efforts resulted in a drastic decrease in malaria cases in Swaziland and South Africa (Sharp et al., 2007; Maharaj et al., 2012).

Examples of cross-border collaborations for infectious disease surveillance and control, which in most cases are not malaria-specific but which could provide models for malaria, include the Connecting Organizations for Regional Disease Surveillance (CORDS), the Middle East Consortium on Infectious Disease Surveillance (MECIDS), the Mekong Basin Disease Surveillance (MBDS), the Asian Partnership on Emerging Infectious Diseases Research, the East African Integrated Disease Surveillance Network, the South African Centre for Disease Surveillance and the South Eastern European Health Network, which links the ministries of health of Albania, Bosnia and Herzegovina, Bulgaria, Croatia, Macedonia, Moldova, Montenegro, Romania and Serbia (Gresham et al., 2011).

The MBDS project, which commenced in 2006 established 16 sites at major border crossings between six countries with the aim to carry out joint, cross-border disease outbreak investigations and responses. The joint team carried out outbreak investigation of malaria between provincial sites in Laos PDR (Savannkhet) and Vietnam (Quang Tri) in 2006 and contained the outbreak (Phommasack et al., 2013). CORDS was established in 2008 and provides a new tool for meeting this social networking challenge on a global scale by fostering the growth of trust-based partnerships among professionals that transcend not just organizational but also geopolitical boundaries (Gresham et al., 2011). MECIDS was established in 2003 and links public health experts and ministry of health officials from Israel, Jordan and Palestine (Gresham et al., 2011). MECIDS played a pivotal role in detecting *salmonella* and mumps outbreaks and containing the influenza viruses H5N1 and H1N1 (Gresham et al., 2009). The strength of such collaboration is prompt sharing of cross-border data. However, there are a number of impediments for such collaborations, including the time taken to build the trust required before cross-border data can be shared freely (Phommasack et al., 2013).

Other collaborations have moved beyond surveillance and disease containment. The Pacific Malaria Initiative was introduced in Vanuatu and Solomon Islands in 2007 to aid in their control efforts with the ultimate goal of malaria elimination. The Asia Pacific Malaria Elimination Network (APMEN) was established in 2009 and represents 15 countries in the Asian Pacific region. The Country Partners, together with regional partners from the academic, development, nongovernmental and private sectors and global agencies, including the WHO, collaboratively address the unique challenges of malaria elimination in the region through leadership, advocacy, capacity building, knowledge exchange and building evidence to support more effective, sustained malaria elimination programmes across the region (APMEN, 2014). Similarly, the Elimination Eight Regional Initiative has been established in Southern Africa to support cross-border collaboration and achievement of mutual goals for malaria elimination in that region. Such international initiatives naturally have a key role to play in developing and implementing strategies to mitigate the threat of cross-border malaria, which is inherently a shared problem between interconnected jurisdictions.

4.2 Surveillance-response and cross-border initiatives

The importance of a robust surveillance-response system at points of entry from areas with local malaria transmission, which facilitates swift treatment and follow-up of infected individuals and their environment, has been recognized (Cohen et al., 2009). Oman has been able to reduce imported cases through mass screening of individuals arriving at the airport from East African countries; those who test positive were treated for free and monitored for two weeks. Both Oman and the UAE have provided free treatment to everyone who tests positive, whether they are nationals or foreigners. Testing for malaria at entry points in Mauritius was shown to provide benefits for investment, by maintaining elimination despite large cyclones in 1994 and 2002 that caused costly damage and an increase in the number of travellers arriving from malaria-endemic countries (Tatarsky et al., 2011; Aboobakar et al., 2012). Screening arriving passengers for malaria at the border points and obtaining a detailed travel history have been deployed to assess the impact of human population movement on malaria in Djibouti (Noor et al., 2011). Proactive prevention programmes to screen all prospective immigrants for malaria infection in their home countries, rather than point of entry, significantly reduced the numbers of imported infections in Kuwait (Iqbal et al., 2003). These approaches work well where border crossings are tightly controlled, but they may be of limited value in remote areas where people pass unchecked between countries.

Population-based surveys that measure cases of parasitaemia can be used to identify high-transmission areas, which often have a low clinical burden because of high rates of immunity in the population. These surveys have the potential to assist malaria control programmes in active detection of transmission hotspots (Clements et al., 2013; Wangdi et al., 2014). However, such surveys become inefficient when malaria incidence is very low because few cases of parasitaemia are identified relative to the sampling effort. Once the interruption of transmission has been achieved in the context of malaria elimination programmes, intensified disease surveillance and swift intervention responses are the basic requirements to prevent the re-establishment of any introduced parasites. In the post-elimination period, the loss of immunity and the high reproductive rate of the malaria parasite in communities where competent vectors are still present could precipitate outbreaks if malaria infections are re-introduced into the population (Greenwood, 2008), emphasising the need for continued surveillance in areas receptive to resurgence.

In some cases, cross-border malaria control, that is, expansion of malaria control programmes from malaria-eliminating countries to neighbouring malaria-controlling countries, might be necessary to create buffering zones to thwart reintroduction of the parasite (Cui et al., 2012). For example, pilot trials of cross-border malaria control at the Thai—Myanmar and Chinese—Myanmar border areas suggest that organized control in these areas is feasible (Richards et al., 2009). Frequent population movements (every three months) across the Thai-Cambodia border and from the border area across Cambodia indicate the need for heightened surveillance for artemisinin resistance outside what has been designated as the containment zone (Khamsiriwatchara et al., 2011). Obviously, such cross-border activities demand coordination of governments between the neighbouring nations. Cross-border dialogue in solving malaria-related issues need to be initiated, and other control and preventive activities such as IRS need to be synchronized to achieve maximum benefits.

Fever surveillance of people who cross border area can be used to identify malaria-associated fever. Offering free treatments would encourage people to avail this service. The GeoSentinel Surveillance in the United States from March 1997 to March 2006 showed that malaria was the most common specific etiologic diagnosis found in 21% of ill returned travellers with fever (Wilson et al., 2007). Kuwait initiated a proactive preventive programme to screen all prospective immigrants for malaria infection in their home countries. As a result, the malaria cases among the immigrants reduced by 52.6% per year (Iqbal et al., 2003). Fever surveillance was mostly used in tourists and travellers from Europe and North America (Wilson et al., 2007; Leder et al., 2004; Journal, 2004). However, it will not be possible to monitor fever, when border crossing takes place through informal border and forest areas. Additionally, fever surveillance would be of limited value in the people who have clinical immunity since these people will not develop fever even when they are infected with *Plasmodia* parasites.

4.3 Strengthening of preventive measures for cross-border malaria

Early diagnosis and prompt treatment of people infected with malaria in malaria-eliminated countries would serve a very important tool in preventing reintroduction. However, to deliver prompt diagnosis and treatment, the health systems in most border areas need to be strengthened. The need to take adequate chemoprophylaxis for people moving from non-endemic to malaria-endemic countries and vice versa is often ignored due to deficient knowledge on the availability of chemoprophylaxis and for financial reasons. Drugs for chemoprophylaxis need to be made available in the border areas. The benefits of sleeping under LLINs need to be highlighted through education, and LLINs being made available through social marketing, as has been done for refugees from Afghanistan in Pakistan (Rowland and Nosten, 2001). Alternative approaches such as the provision of insecticide-treated hammocks for people frequenting forest areas in border areas (Magris et al., 2007; Thang et al., 2009), deltamethrin-sprayed tarpaulins or tents, and premethrin-treated blankets and top sheets provide more promising options for people overnighting in the forest (Graham et al., 2002; Rowland and Nosten, 2001). These protective measures achieve the goal of reducing exposure to infected vectors for populations who do not live in traditional housing each night, a feature common to people in border areas.

4.4 Technological solutions to support operational decision making and surveillance response

Spatial decision support system (SDSS) provide enhanced support for decision making, and management using data that has spatial or geographical components (Keenan, 2003). SDSS are generally based on a database housed within a geographic information system with an interactive mapping interface. SDSS can contain modules for planning, monitoring and evaluating the delivery and coverage of interventions including IRS and LLIN within target populations, and for mapping malaria surveillance data (Kelly et al., 2013, 2011; Reid et al., 2010; Srivastava et al., 2009; Zhang et al., 2008).

Erhart, A., Thang, N.D., Xa, N.X., Thieu, N.Q., Hung, L.X., Hung, N.Q., Nam, N.V., Yoi, L.V., Tung, N.M., Bien, T.H., Tuy, T.Q., Cong, L.D., Thuan, L.K., Coosemans, M., D'Alessandro, U., 2007. Accuracy of the health information system on malaria surveillance in Vietnam. Trans. R. Soc. Trop. Med. Hyg. 101, 216—225.

Feachem, R., Sabot, O., 2008. A new global malaria eradication strategy. Lancet 371, 1633—1635.

Feachem, R.G., Phillips, A.A., Hwang, J., Cotter, C., Wielgosz, B., Greenwood, B.M., Sabot, O., Rodriguez, M.H., Abeyasinghe, R.R., Ghebreyesus, T.A., Snow, R.W., 2010a. Shrinking the malaria map: progress and prospects. Lancet 376, 1566—1578.

Feachem, R.G., Phillips, A.A., Targett, G.A., Snow, R.W., 2010b. Call to action: priorities for malaria elimination. Lancet 376, 1517—1521.

Fenner, L., Weber, R., Steffen, R., Schlagenhauf, P., 2007. Imported infectious disease and purpose of travel, Switzerland. Emerg. Infect. Dis. 13, 217—222.

Froude, J.R., Weiss, L.M., Tanowitz, H.B., Wittner, M., 1992. Imported malaria in the Bronx: review of 51 cases recorded from 1986 to 1991. Clin. Infect. Dis. 15, 774—780.

Fulford, M., Keystone, J.S., 2005. Health risks associated with visiting friends and relatives in developing countries. Curr. Infect. Dis. Rep. 7, 48—53.

Gamage-Mendis, A.C., Carter, R., Mendis, C., De Zoysa, A.P., Herath, P.R., Mendis, K.N., 1991. Clustering of malaria infections within an endemic population: risk of malaria associated with the type of housing construction. Am. J. Trop. Med. Hyg. 45, 77—85.

Gonzalez, M.C., Hidalgo, C.A., Barabasi, A.-L., 2008. Understanding individual human mobility patterns. Nature 453, 779—782.

Graham, K., Mohammad, N., Rehman, H., Nazari, A., Ahmad, M., Kamal, M., Skovmand, O., Guillet, P., Allan, R., Zaim, M., Yates, A., Lines, J., Rowland, M., 2002. Insecticide-treated plastic tarpaulins for control of malaria vectors in refugee camps. Med. Vet. Entomol. 16, 404—408.

Gratz, N.G., 1999. Emerging and resurging vector-borne diseases. Annu. Rev. Entomol. 44, 51—75.

Greenwood, B.M., 2008. Control to elimination: implications for malaria research. Trends Parasitol. 24, 449—454.

Gresham, L., Ramlawi, A., Briski, J., Richardson, M., Taylor, T., 2009. Trust across borders: responding to 2009 H1N1 influenza in the Middle East. Biosecur. Bioterr. 7, 399—404.

Gresham, L.S., Pray, L.A., Wibulpolprasert, S., Trayner, B., 2011. Public—private partnerships in trust-based public health social networking: connecting organizations for regional disease surveillance (CORDS). J. Commer. Biotec. 17, 241—247.

Guerra, C., Snow, R., Hay, S., 2006. A global assessment of closed forests, deforestation and malaria risk. Ann. Trop. Med. Parasitol. 100, 189.

Gueye, C., Teng, A., Kinyua, K., Wafula, F., Gosling, R., McCoy, D., 2012. Parasites and vectors carry no passport: how to fund cross-border and regional efforts to achieve malaria elimination. Malar. J. 11, 344.

Hira, P.R., Behbehani, K., Al-Kandari, S., 1985. Imported malaria in Kuwait. Trans. R. Soc. Trop. Med. Hyg. 79, 291—296.

Hira, P.R., Al-Ali, F., Soriano, E.B., Behbehani, K., 1988. Aspects of imported malaria at a district general hospital in non-endemic Kuwait, Arabian Gulf. Eur. J. Epidemiol. 4, 200—205.

Hiwat, H., Hardjopawiro, L., Takken, W., Villegas, L., 2012. Novel strategies lead to pre-elimination of malaria in previously high-risk areas in Suriname, South America. Malar. J. 11, 10.

Hoyer, S., Nguon, S., Kim, S., Habib, N., Khim, N., Sum, S., Christophel, E.-M., Bjorge, S., Thomson, A., Kheng, S., 2012. Focused Screening and Treatment (FSAT): a PCR-based strategy to detect malaria parasite carriers and contain drug resistant *P. falciparum*, Pailin, Cambodia. PLoS One 7, e45797.

Hu, H., Singhasivanon, P., Salazar, N.P., Thimasarn, K., Li, X., Wu, Y., Yang, H., Zhu, D., Supavej, S., Looarecsuwan, S., 1998. Factors influencing malaria endemicity in Yunnan Province, PR China (analysis of spatial pattern by GIS). Geographical Information System. Southeast Asian J. Trop. Med. Public Health 29, 191–200.

Huguet, J., Punpuing, S., 2005. International Migration in Thailand. International Organization for Migration. Regional Office Bangkok, Thailand.

Hung le, Q., Vries, P.J., Giao, P.T., Nam, N.V., Binh, T.Q., Chong, M.T., Quoc, N.T., Thanh, T.N., Hung, L.N., Kager, P.A., 2002. Control of malaria: a successful experience from Viet Nam. Bull. World Health Organ. 80, 660–666.

Iqbal, J., Hira, P.R., Al-Ali, F., Sher, A., 2003. Imported malaria in Kuwait (1985–2000). J. Travel Med. 10, 324–329.

Jackson, S., Sleigh, A., 2000. Resettlement for China's Three Gorges Dam: socio-economic impact and institutional tensions. Communist Post-Communist Stud. 33, 223–241.

Journal, M., 2004. Epidemiology and clinical features of *vivax* malaria imported to Europe: sentinel surveillance data from TropNetEurop. Malar. J. 3.

Kansakar, V.B.S., 2001. Nepal–India Open Border: Prospects, Problems and Challenges. Nepal Democracy. hp://www.nepaldemocracy.org.

Keenan, P.B., 2003. Spatial Decision Support Systems. Decision Making Support Systems: Achievements and Challenges for the New Decade, pp. 28–39.

Keiser, J., De Castro, M.C., Maltese, M.F., Bos, R., Tanner, M., Singer, B.H., Utzinger, J., 2005. Effect of irrigation and large dams on the burden of malaria on a global and regional scale. Am. J. Trop. Med. Hyg. 72, 392–406.

Kelly, G.C., Hii, J., Batarii, W., Donald, W., Hale, E., Nausien, J., Pontifex, S., Vallely, A., Tanner, M., Clements, A., 2010. Modern geographical reconnaissance of target populations in malaria elimination zones. Malar. J. 9, 289.

Kelly, G.C., Seng, C.M., Donald, W., Taleo, G., Nausien, J., Batarii, W., Iata, H., Tanner, M., Vestergaard, L.S., Clements, A.C., 2011. A spatial decision support system for guiding focal indoor residual interventions in a malaria elimination zone. Geospat. Health 6, 21–31.

Kelly, G.C., Hale, E., Donald, W., Batarii, W., Bugoro, H., Nausien, J., Smale, J., Palmer, K., Bobogare, A., Taleo, G., Vallely, A., Tanner, M., Vestergaard, L.S., Clements, A.C., 2013. A high-resolution geospatial surveillance-response system for malaria elimination in Solomon Islands and Vanuatu. Malar. J. 12, 108.

Khamsiriwatchara, A., Wangroongsarb, P., Thwing, J., Eliades, J., Satimai, W., Delacollette, C., Kaewkungwal, J., 2011. Respondent-driven sampling on the Thailand–Cambodia border. I. Can malaria cases be contained in mobile migrant workers? Malar. J. 10, 120.

Kitvatanachai, S., Janyapoon, K., Rhongbutsri, P., Thap, L.C., 2003. A survey on malaria in mobile Cambodians in Aranyaprathet, Sa Kaeo Province, Thailand. Southeast Asian J. Trop. Med. Public Health 34, 48–53.

Koita, K., Novotny, J., Kunene, S., Zulu, Z., Ntshalintshali, N., Gandhi, M., Gosling, R., 2013. Targeting imported malaria through social networks: a potential strategy for malaria elimination in Swaziland. Malar. J. 12, 219.

Konradsen, F., Amerasinghe, P., Van der hoek, W., Amerasinghe, F., Perera, D., Piyaratne, M., 2003. Strong association between house characteristics and malaria vectors in Sri Lanka. Am. J. Trop. Med. Hyg. 68, 177—181.

Koyadun, S., Bhumiratana, A., 2005. Surveillance of imported bancroftian filariasis after two-year multiple-dose diethylcarbamazine treatment. Southeast Asian J. Trop. Med. Public Health 36, 822—831.

Leder, K., Black, J., O'Brien, D., Greenwood, Z., Kain, K.C., Schwartz, E., Brown, G., Torresi, J., 2004. Malaria in travelers: a review of the GeoSentinel surveillance network. Clin. Infect. Dis. 39, 1104—1112.

Lee, T.J., Mullany, L.C., Richards, A.K., Kuiper, H.K., Maung, C., Beyrer, C., 2006. Mortality rates in conflict zones in Karen, Karenni, and Mon states in eastern Burma. Trop. Med. Int. Health 11, 1119—1127.

Lindsay, S.W., Emerson, P.M., Charlwood, J.D., 2002. Reducing malaria by mosquito-proofing houses. Trends Parasitol. 18, 510—514.

Liu, Y., Hsiang, M.S., Zhou, H., Wang, W., Cao, Y., Gosling, R.D., Cao, J., Gao, Q., 2014. Malaria in overseas labourers returning to China: an analysis of imported malaria in Jiangsu Province, 2001—2011. Malar. J. 13, 29.

Magris, M., Rubio-Palis, Y., Alexander, N., Ruiz, B., Galvan, N., Frias, D., Blanco, M., Lines, J., 2007. Community-randomized trial of lambdacyhalothrin-treated hammock nets for malaria control in Yanomami communities in the Amazon region of Venezuela. Trop. Med. Int. Health 12, 392—403.

Maharaj, R., Morris, N., Seocharan, I., Kruger, P., Moonasar, D., Mabuza, A., Raswiswi, E., Raman, J., 2012. The feasibility of malaria elimination in South Africa. Malar. J. 11, 423.

malERA Consultative Group on Monitoring, Evaluation, and Surveillance, 2011. A research agenda for malaria eradication: monitoring, evaluation, and surveillance. PLoS Med. 8, e1000400.

Manh, B.H., Clements, A.C., Thieu, N.Q., Hung, N.M., Hung, L.X., Hay, S.I., Hien, T.T., Wertheim, H.F., Snow, R.W., Horby, P., 2011. Social and environmental determinants of malaria in space and time in Viet Nam. Int. J. Parasitol. 41, 109—116.

Martens, P., Hall, L., 2000. Malaria on the move: human population movement and malaria transmission. Emerg. Infect. Dis. 6, 103—109.

Mendis, K., Rietveld, A., Warsame, M., Bosman, A., Greenwood, B., Wernsdorfer, W.H., 2009. From malaria control to eradication: the WHO perspective. Trop. Med. Int. Health 14, 802—809.

Mittal, P.K., Wijeyaratne, P., Pandey, S., 2004. Status of insecticide resistance of malaria, Kala-azar and Japanese encephalitis vectors in Bangladesh, Bhutan, India and Nepal (BBIN). Environ. Health Proj. Act. Rep. 129, 44—48.

Muentener, P., Schlagenhauf, P., Steffen, R., 1999. Imported malaria (1985—95): trends and perspectives. Bull. World Health Organ. 77, 560—566.

Murray, C.J.L., Rosenfeld, L.C., Lim, S.S., Andrews, K.G., Foreman, K.J., Haring, D., Fullman, N., Naghavi, M., Lozano, R., Lopez, A.D., 2012. Global malaria mortality between 1980 and 2010: a systematic analysis. Lancet 379, 413—431.

Najera, J., Koumetsov, R., Delacollette, C., 1998. Malaria Epidemics Detection and Control Forescating and Prevention. WHO.

Noor, A., Mohamed, M., Mugyenyi, C., Osman, M., Guessod, H., Kabaria, C., Ahmed, I., Nyonda, M., Cook, J., Drakeley, C., Mackinnon, M., Snow, R., 2011. Establishing the extent of malaria transmission and challenges facing pre-elimination in the Republic of Djibouti. BMC Infect. Dis. 11, 121.

Noor, A., Uusiku, P., Kamwi, R., Katokele, S., Ntomwa, B., Alegana, V., Snow, R., 2013. The receptive versus current risks of *Plasmodium falciparum* transmission in Northern Namibia: implications for elimination. BMC Infect. Dis. 13, 184.

Olson, S.H., Gangnon, R., Silveira, G.A., Patz, J.A., 2010. Deforestation and malaria in Mancio Lima County, Brazil. Emerg. Infect. Dis. 16, 1108−1115.

Packard, R.M., 1986. Agricultural development, migrant labor and the resurgence of malaria in Swaziland. Soc. Sci. Med. 22, 861−867.

Patz, J.A., Graczyk, T.K., Geller, N., Vittor, A.Y., 2000. Effects of environmental change on emerging parasitic diseases. Int. J. Parasitol. 30, 1395−1405.

Phommasack, B., Jiraphongsa, C., Ko Oo, M., Bond, K.C., Phaholyothin, N., Suphanchaimat, R., Ungchusak, K., Macfarlane, S.B., 2013. Mekong Basin Disease Surveillance (MBDS): a trust-based network. Emerg. Health Threats J. 6.

Pichainarong, N., Chaveepojnkamjorn, W., 2004. Malaria infection and life-style factors among hilltribes along the Thai-Myanmar border area, northern Thailand. Southeast Asian J. Trop. Med. Public Health 35, 834−839.

Pindolia, D., Garcia, A., Wesolowski, A., Smith, D., Buckee, C., Noor, A., Snow, R., Tatem, A., 2012. Human movement data for malaria control and elimination strategic planning. Malar. J. 11, 205.

Pongvongsa, T., Ha, H., Thanh, L., Marchand, R., Nonaka, D., Tojo, B., Phongmany, P., Moji, K., Kobayashi, J., 2012. Joint malaria surveys lead towards improved cross-border cooperation between Savannakhet province, Laos and Quang Tri province, Vietnam. Malar. J. 11, 262.

Prothero, R.M., 1977. Disease and mobility: a neglected factor in epidemiology. Int. J. Epidemiol. 6, 259−267.

Prothero, R.M., 1995. Malaria in Latin America: environmental and human factors. Bull. Lat. Am. Res. 14, 357−365.

Prothero, R.M., 1999. Malaria, forests and people in Southeast Asia. Singap. J. Trop. Geogr. 20, 76−85.

Reid, H., Vallely, A., Taleo, G., Tatem, A.J., Kelly, G., Riley, I., Harris, I., Henri, I., Iamaher, S., Clements, A.C., 2010. Baseline spatial distribution of malaria prior to an elimination programme in Vanuatu. Malar. J. 9, 150.

Reza, S., Abbas, M., Massoud, H., Aliakbar, S., Fatemeh, S., 2009. Epidemiology of malaria in Khorasan Razavi province north-eastern of Iran within last 7 years (April 2001−March 2008). Int. J. Parasit. Dis. 4.

Richards, A.K., Banek, K., Mullany, L.C., Lee, C.I., Smith, L., Oo, E.K., Lee, T.J., 2009. Cross-border malaria control for internally displaced persons: observational results from a pilot programme in eastern Burma/Myanmar. Trop. Med. Int. Health 14, 512−521.

Roll Back Malaria, 2008. The Global Malaria Action Plan, for a Malaria-Free World. http://www.rbm.who.int/gmap/. downloaded on 29/10/2014.

Roseman, C.C., 1971. Migration as a spatial and temporal process. Ann. Assoc. Am. Geogr. 61, 589−598.

Rowland, M., Nosten, F., 2001. Malaria epidemiology and control in refugee camps and complex emergencies. Ann. Trop. Med. Parasitol. 95, 741−754.

Satitvipawee, P., Wongkhang, W., Pattanasin, S., Hoithong, P., Bhumiratana, A., 2012. Predictors of malaria-association with rubber plantations in Thailand. BMC Public Health 12, 1115.

Schlagenhauf, P., Steffen, R., Loutan, L., 2003. Migrants as a major risk group for imported malaria in European countries. J. Travel Med. 10, 106−107.

Schultz, M.G., 1989. Malaria in migrants and travellers. Trans. R. Soc. Trop. Med. Hyg. 83 (Suppl.), 31—34.
Scolari, C., Tedoldi, S., Casalini, C., Scarcella, C., Matteelli, A., Casari, S., El Hamad, I., Castelli, F., 2002. Knowledge, attitudes, and practices on malaria preventive measures of migrants attending a public health clinic in northern Italy. J. Travel Med. 9, 160—162.
Sharp, B.L., Kleinschmidt, I., Streat, E., Maharaj, R., Barnes, K.I., Durrheim, D.N., Ridl, F.C., Morris, N., Seocharan, I., Kunene, S., La Grange, J.J.P., Mthembu, J.D., Maartens, F., Martin, C.L., Barreto, A., 2007. Seven years of regional malaria control collaboration—Mozambique, South Africa, and Swaziland. Am. J. Trop. Med. Hyg. 76, 42—47.
Singh, N., Mehra, R.K., Sharma, V.P., 1999. Malaria and the Narmada-river development in India: a case study of the Bargi dam. Ann. Trop. Med. Parasitol. 93, 477—488.
Singh, N., Mishra, A.K., 2000. Anopheline ecology and malaria transmission at a new irrigation project area (Bargi Dam) in Jabalpur (Central India). J. Am. Mosq. Control Assoc. 16, 279—287.
Srivastava, A., Nagpal, B.N., Joshi, P.L., Paliwal, J.C., Dash, A.P., 2009. Identification of malaria hot spots for focused intervention in tribal state of India: a GIS based approach. Int. J. Health Geogr. 8, 30.
Stager, K., Legros, F., Krause, G., Low, N., Bradley, D., Desai, M., Graf, S., D'Amato, S., Mizuno, Y., Janzon, R., Petersen, E., Kester, J., Steffen, R., Schlagenhauf, P., 2009. Imported malaria in children in industrialized countries, 1992—2002. Emerg. Infect. Dis. 15, 185—191.
Stern, A., 1998. International population movements and public health in the Mekong region: an overview of some issues concerning mapping. Southeast Asian J. Trop. Med. Public Health 29, 201—212.
Stoddard, S.T., Morrison, A.C., Vazquez-Prokopec, G.M., Paz Soldan, V., Kochel, T.J., Kitron, U., Elder, J.P., Scott, T.W., 2009. The role of human movement in the transmission of vector-borne pathogens. PLoS Negl. Trop. Dis. 3, e481.
Sultan, D.M., Khalil, M.M., Abdouh, A.S., Doleh, W.F., Al Muthanna, A.A., 2009. Imported malaria in United Arab Emirates: evaluation of a new DNA extraction technique using nested PCR. Korean J. Parasitol. 47, 227—233.
Tatarsky, A., Aboobakar, S., Cohen, J.M., Gopee, N., Bheecarry, A., Moonasar, D., Phillips, A.A., Kahn, J.G., Moonen, B., Smith, D.L., Sabot, O., 2011. Preventing the reintroduction of malaria in Mauritius: a programmatic and financial assessment. PLoS One 6, e23832.
Tatem, A.J., Qiu, Y., Smith, D.L., Sabot, O., Ali, A.S., Moonen, B., 2009. The use of mobile phone data for the estimation of the travel patterns and imported *Plasmodium falciparum* rates among Zanzibar residents. Malar. J. 8, 287.
Tatem, A.J., Smith, D.L., 2010. International population movements and regional *Plasmodium falciparum* malaria elimination strategies. Proc. Natl. Acad. Sci. U.S.A. 107, 12222—12227.
Thang, N.D., Erhart, A., Speybroeck, N., Xa, N.X., Thanh, N.N., Van Ky, P., Hung, L.X., Coosemans, M., D'Alessandro, U., 2009. Long-lasting insecticidal hammocks for controlling forest malaria: a community-based trial in a rural area of central Vietnam. PLoS One 4, e7369.
The World Bank, 2012. Fragile and Conflict Affected Situations. The World Bank.
Thimasarn, K., 2003. A Strategic Framework for Rolling Back Malaria in the Mekong Region (Mimeographed Document).

Vittor, A.Y., Gilman, R.H., Tielsch, J., Glass, G., Shields, T., Lozano, W.S., Pinedo-Cancino, V., Patz, J.A., 2006. The effect of deforestation on the human-biting rate of *Anopheles darlingi*, the primary vector of *falciparum* malaria in the Peruvian Amazon. Am. J. Trop. Med. Hyg. 74, 3–11.

Vittor, A.Y., Pan, W., Gilman, R.H., Tielsch, J., Glass, G., Shields, T., Sanchez-Lozano, W., Pinedo, V.V., Salas-Cobos, E., Flores, S., Patz, J.A., 2009. Linking deforestation to malaria in the Amazon: characterization of the breeding habitat of the principal malaria vector, *Anopheles darlingi*. Am. J. Trop. Med. Hyg. 81, 5–12.

Wagner, K.S., Lawrence, J., Anderson, L., Yin, Z., Delpech, V., Chiodini, P.L., Redman, C., Jones, J., 2013. Migrant health and infectious diseases in the UK: findings from the last 10 years of surveillance. J. Public Health (Oxf) 36, 28–35.

Walsh, J.F., Molyneux, D.H., Birley, M.H., 1993. Deforestation: effects on vector-borne disease. Parasitology 106, 55–75.

Wangdi, K., Singhasivanon, P., Silawan, T., Lawpoolsri, S., White, N.J., Kaewkungwal, J., 2010. Development of temporal modelling for forecasting and prediction of malaria infections using time-series and ARIMAX analyses: a case study in endemic districts of Bhutan. Malar. J. 9, 251.

Wangdi, K., Kaewkungwal, J., Singhasivanon, P., Silawan, T., Lawpoolsri, S., White, N.J., 2011. Spatio-temporal patterns of malaria infection in Bhutan: a country embarking on malaria elimination. Malar. J. 10, 89.

Wangdi, K., Gatton, M., Kelly, G., Clements, A., 2014. Prevalence of asymptomatic malaria and bed net ownership and use in Bhutan, 2013: a country earmarked for malaria elimination. Malar. J. 13, 352.

Wangroongsarb, P., Sudathip, P., Satimai, W., 2012. Characteristics and malaria prevalence of migrant populations in malaria-endemic areas along the Thai–Cambodian border. Southeast Asian J. Trop. Med. Public Health 43, 261–269.

Wijeyaratne, P.M., Chand, P.B., Valecha, N., Shahi, B., Adak, T., Ansari, M.A., Jha, J., Pandey, S., Bannerjee, S., Bista, M.B., 2005. Therapeutic efficacy of antimalarial drugs along the eastern Indo-Nepal border: a cross-border collaborative study. Trans. R. Soc. Trop. Med. Hyg. 99, 423–429.

Williams, H., Hering, H., Spiegel, P., 2013. Discourse on malaria elimination: where do forcibly displaced persons fit in these discussions? Malar. J. 12, 121.

Wilson, M.E., Weld, L.H., Boggild, A., Keystone, J.S., Kain, K.C., von Sonnenburg, F., Schwartz, E., Network, G.S., 2007. Fever in returned travelers: results from the GeoSentinel surveillance network. Clin. Infect. Dis. 44, 1560–1568.

Wisit Chaveepojnkamjorn, D., 2005. Behavioral factors and malaria infection among the migrant population, Chiang Rai province. J. Med. Assoc. Thai. 88, 1293–1301.

Wolpert, J., 1965. Behavioral aspects of the decision to migrate. Pap. Reg. Sci. 15, 159–169.

WHO, 2010. Malaria in the Mekong Subregion: Regional and Country Profiles. World Health Organization, New Delhi, India.

WHO, 2012. World Malaria Report 2011. World Health Organization, Geneva.

WHO, 2013. World Malaria Report 2012. World Health Organization, Geneva.

Xu, J., Liu, H., 1997. Border malaria in Yunnan, China. Southeast Asian J. Trop. Med. Public Health 28, 456–459.

Xu, J., Liu, H., 2012. The challenges of malaria elimination in Yunnan Province, People's Republic of China. Southeast Asian J. Trop. Med. Public Health 43, 819–824.

Yangzom, T., Gueye, C., Namgay, R., Galappaththy, G., Thimasarn, K., Gosling, R., Murugasampillay, S., Dev, V., 2012. Malaria control in Bhutan: case study of a country embarking on elimination. Malar. J. 11, 9.

Zhang, W., Wang, L., Fang, L., Ma, J., Xu, Y., Jiang, J., Hui, F., Wang, J., Liang, S., Yang, H., 2008. Spatial analysis of malaria in Anhui province, China. Malar. J. 7, 19.

CHAPTER THREE

Development of Malaria Transmission-Blocking Vaccines: From Concept to Product

Yimin Wu[*,†,1], Robert E. Sinden[§], Thomas S. Churcher[¶], Takafumi Tsuboi[‖], Vidadi Yusibov[#]

[*]Laboratory of Malaria Immunology and Vaccinology, National Institute of Allergy and Infectious Diseases, Rockville, MD, USA
[§]The Jenner Institute, Oxford, UK
[¶]MRC Centre for Outbreak Analysis and Modelling, Department of Infectious Disease Epidemiology, School of Public Health, Imperial College London, London, UK
[‖]Division of Malaria Research, Ehime University, Matsuyama, Ehime, Japan
[#]Fraunhofer USA Center for Molecular Biotechnology, Newark, DE, USA
[†]Current affiliation: PATH-Malaria Vaccine Initiative, Washington DC, USA
[1]Corresponding author: E-mail: ywu@path.org

Contents

1. Introduction	110
2. Targets for TBVs	113
2.1 Parasite surface: pre-activation and post-activation targets	113
2.1.1 Structural features, spatial and temporal expression of parasite targets	113
2.1.2 Sequence polymorphism of parasite targets in field isolates	116
2.2 Ookinete-secreted proteins	117
2.3 Mosquito components	119
3. Vaccine Development Efforts and Status	121
3.1 Challenge in production of recombinant vaccines	122
3.2 Challenges in vaccine formulation	126
3.3 Alternative approaches to enhance immunogenicity of candidate vaccines	128
3.4 Multicomponent vaccines	130
4. Protective Correlates and Surrogate Assays for Efficacy	131
4.1 Antibody titre	132
4.2 Mosquito infection in laboratory feeding assays	132
4.3 Laboratory-based population assay	136
4.3.1 TPP and the path forward	137
Acknowledgements	138
References	138

Abstract

Despite decades of effort battling against malaria, the disease is still a major cause of morbidity and mortality. Transmission-blocking vaccines (TBVs) that target sexual stage parasite development could be an integral part of measures for malaria elimination. In the 1950s, Huff et al. first demonstrated the induction of transmission-blocking immunity in chickens by repeated immunizations with *Plasmodium gallinaceum*-infected red blood cells. Since then, significant progress has been made in identification of parasite antigens responsible for transmission-blocking activity. Recombinant technologies accelerated evaluation of these antigens as vaccine candidates, and it is possible to induce effective transmission-blocking immunity in humans both by natural infection and now by immunization with recombinant vaccines. This chapter reviews the efforts to produce TBVs, summarizes the current status and advances and discusses the remaining challenges and approaches.

1. INTRODUCTION

The goal to eliminate and eradicate malaria set at the Gates Malaria Forum 2007 (Roberts and Enserink, 2007) stimulated political commitment and financial investment to fight malaria, a human disease caused by parasites of *Plasmodium* spp. and transmitted by *Anopheles* mosquitoes. Morbidity and mortality caused by malaria has recently decreased; however, the disease still claimed 584,000 lives in 2013 (WHO Malaria Report, 2014). While a partially protective vaccine (RTS,S) is likely to be licensed within a few years (Moorthy et al., 2013), both the level and duration of protection of that vaccine are currently suboptimal, and RTS,S reportedly acts by reducing clinical disease rather than by preventing infection (Agnandji et al., 2011; Olotu et al., 2013). A vaccine that interrupts malaria transmission would be a valuable additional resource in the fight to eliminate and eradicate the disease.

The complete life cycle of malaria parasites requires both a mosquito host where the parasite undergoes sexual reproduction and a human host where significant asexual replication occurs. Transmission from humans to mosquitoes begins when a mosquito ingests a blood meal from a malaria-infected host. Inside the mosquito midgut and enclosed by a newly synthesized peritrophic matrix (PM), the blood meal contains erythrocytes infected with parasites, including a small proportion of male and female gametocytes. Triggered by a drop in environmental temperature, a rise in pH, and notably by mosquito-derived xathurenic acid, the intracellular gametocytes egress from within erythrocytes, the male gametocytes exflagellate and fertilize the female gametes. During the next 24 h the resulting zygotes undergo further development and differentiation to become banana shaped, motile ookinetes. Subsequently ookinetes penetrate the PM and the midgut epithelium, reach the basal lamina and differentiate to oocysts. Within an oocyst multiple mitotic divisions over the following 7–15 days may result in tens of thousands of sporozoites which, upon maturation, are pumped through the aorta from the ruptured oocyst around the haemocoel from where they invade the salivary glands. The sporozoites are injected into humans when the infected mosquito probes and takes another blood meal. Following inoculation, sporozoites are present in the blood stream very briefly before entering the liver and invading hepatocytes where multiple mitotic divisions can result in tens of thousands of merozoites. Upon maturation these merozoites are released into the blood stream, beginning repeated cycles of invasion of erythrocytes, mitotic division and release of newly formed merozoites. A small proportion (5–20%) of the merozoites leave this cycle of erythrocytic asexual replication and undergo sexual development differentiating into male and female gametocytes which can then be transmitted to mosquitoes in the blood meal.

Figure 1 indicates stages of malaria life cycle that can be targeted for transmission blocking. Among these stages, sexual development in mosquito midgut might represent the most vulnerable target for blocking of malaria transmission. First, the largely intracellular malaria parasite becomes 'naked' upon entering mosquito midgut and remains accessible to intervening agents for up to 24 h. Second, a significant reduction in number occurs during each developmental succession from gametocytes, gametes, zygotes, ookinetes

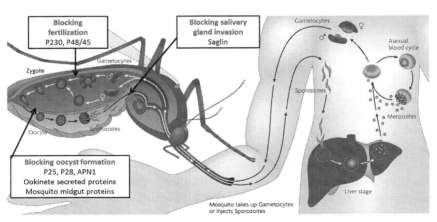

Figure 1 Malaria life cycle *(Revised with permission Su et al. (2007).)* and transmission-blocking targets at (1) fertilization of male and female gametes to form zygote (P230 and P48/45); (2) development of ookinetes, penetration of peritrophic matrix and midgut epithelium to form oocysts at the basal face of the midgut (P25, P28, Chitnase, WARP, CTRP, MAOP, SOAP, CelTos, APN1, CBP, Serpin) and (3) invasion of salivary glands by sporozoites (saglin).

and finally oocysts (Alavi et al., 2003; Vaughan et al., 1992), indicating this as a biological bottleneck during parasite development (Sinden, 2010). The mosquito-stage transmission-blocking vaccines (TBVs), the focus of this chapter, are designed to induce antibody responses in vaccinees. When ingested into blood meals by mosquitoes these antibodies impair parasite viability, inhibit parasite development or interfere with parasite−midgut interaction, thereby reducing or blocking malaria transmission to the mosquito and thence other human hosts.

The induction of transmission-reducing antibodies in a vertebrate host was first demonstrated by Huff et al. (1958) in a bird malaria model. By repeated immunization of avian hosts with killed, mixed-bloodstage *Plasmodium fallax* or *Plasmodium gallinaceum* prior to infecting the birds with live parasites, they observed a reduction in oocyst formation in *Aedes* mosquitoes. Using highly purified gametocytes of *P. gallinaceum*, Gwadz and Carter induced transmission-reducing antibodies by repeated immunization of birds, and the induced antibodies recognized gametocyte proteins (Carter and Chen, 1976; Gwadz, 1976). These proteins were later characterized as P230, P48/45, P28 and P25 (Kaushal et al., 1983) and were expressed on the surface of gametes, zygotes or ookinetes (Carter and Kaushal, 1984; Kumar and Carter, 1985; Vermeulen et al., 1985a). Monoclonal antibodies against some of these proteins reduced or blocked *P. gallinaceum* infectivity to mosquitoes in an ex vivo assay in which mosquitoes were fed through a membrane feeder on a parasite culture mixed with test antibodies (Grotendorst et al., 1984; Kaushal et al., 1983).

The design and development of TBVs are affected by several unique features of the vaccines. (1) The vaccine is designed to work outside the vaccinee through antibody-mediated activities. This feature allows vaccine development efforts uniquely to focus on inducing adequate humoral responses. (2) Vaccine targets could be mosquito moieties or parasite components expressed in mosquito. In some cases, the absence of expression of the target antigen in human host correlates with an observed lack of genetic polymorphism presumably induced by immune pressure in humans, therefore vaccine design might be based one parasite strain yet induce strain-transcending immunity. On the other hand, absence of the target antigens in the human host during natural infection excludes potential boosting effects by natural malaria infection in areas where infection rates remain high. Thus vaccine development efforts necessarily aim to sustain an adequate level of antibodies for an adequate duration, ideally more than one transmission season. (3) The vaccine does not immediately reduce the chance of malaria infection in the vaccinated individuals. Rather, vaccinees may be protected through herd immunity. Whereas herd immunity has been demonstrated to play important roles in preventing spread of multiple diseases (Buttery et al., 2011; Gastanaduy et al., 2013; Rashid et al., 2012; Waye et al., 2013), there is no precedent for licensure of a vaccine overtly based on herd protection.

2. TARGETS FOR TBVs

The means for identifying TBV candidates varied from recognition by sexual stage-specific antibodies to structural resemblance to proteins of other organisms with known biological activities. Confirmation of biological function of the candidates, however, almost invariably involves an ex vivo assay in which a gametocyte culture is reconstituted in blood- and serum-containing antibodies raised against the candidates, offered to mosquitoes from a laboratory colony in a membrane feeder behind an artificial membrane through which the mosquitoes feed. 7 to 8 days post feeds mosquitoes are dissected and oocysts in the midgut are enumerated. Transmission-blocking activity (TBA) and transmission-reducing activity (TRA) are quantified by reduction, compared to a control feed in which the blood meal contains control antibodies, in infection prevalence or in infection intensity, respectively. The assay, termed standard membrane feeding assay (SMFA), is considered by some to be a 'gold standard' for assessing the interference of mosquito-to-human transmission, and will be discussed in detail in Section 4.2.

2.1 Parasite surface: pre-activation and post-activation targets

2.1.1 Structural features, spatial and temporal expression of parasite targets

Homologues of the surface antigens originally found in *P. gallinaceum* were also identified in human malaria parasite *Plasmodium falciparum*, prior to activation in gametocytes and following activation in gametes, zygotes and ookinetes, using transmission-blocking monoclonal antibodies raised against purified *P. falciparum* macrogametes (Carter et al., 1990; Quakyi et al., 1987; Read et al., 1994; Roeffen et al., 2001; Vermeulen et al., 1985a, 1986). These include: Pfs230, a protein on the surface of freshly emerged (activated) gametes and so named for its apparent molecular weight by SDS-PAGE; Pfs48/45, another surface protein of gametes and so named for its apparent 48 and 45 kDa doublet on SDS-PAGE (Kumar and

Carter, 1984; Rener et al., 1983); and Pfs25, a 25 kDa protein expressed on surface of macrogametes, zygotes and ookinetes (Vermeulen et al., 1985b). Pfs28, a paralog of Pfs25 and expressed on the ookinete surface, is also a potential target of transmission-reducing activities (Duffy and Kaslow, 1997). Orthologous genes of these proteins have been found in other *Plasmodium* spp., including *Plasmodium vivax*, *Plasmodium berghei* and *Plasmodium yoelii*, and the products are collectively called P230, P48/45, P25 and P28. Based on their temporal expression on the parasite surface, P230 and P48/45 are known as pre-activation targets and P25 and P28, post-activation targets. Subsequent research on these proteins in *P. vivax*, another human malaria parasite, has underpinned the development of anti-*P. vivax* TBVs.

P230 belongs to a structural family typified by partially conserved cysteine motifs paired as cysteine-rich double domains highly constrained by disulfide bonds (Carter et al., 1995; Doi et al., 2011; Gerloff et al., 2005; Templeton and Kaslow, 1999). The full length precursor of Pfs230 which is detected in gametocytes has molecular weight of 360 kDa, contains 14 cysteine-rich motifs, each containing 4–6 cysteines, paired as 7 double domains and an N-terminal pro-domain featuring a signal sequence, a glutamic acid rich region including 25-tandem glutamates and 16 tandem copies of degenerate E E/G V/E G repeats. Upon emergence of the gametes from the enveloping erythrocyte membrane the pro-domain is cleaved, resulting in mature proteins of 307 and 300 kDa, each retaining the 14 cysteine motifs being located on the surface of gametes (Williamson et al., 1993, 1996). Mature Pfs230 has no transmembrane domain or other membrane anchor motif and reportedly attaches to the gamete surface through interaction with Pfs48/45 (Kumar, 1985). Deleting partial or entire cysteine motifs from Pfs230 did not affect emergence of morphologically normal gametes; however, the truncated Pfs230s were no longer retained on the surface of gametes (Eksi et al., 2002). Moreover, the emerged microgametes expressing truncated Pfs230 failed to attach to erythrocytes and form exflagellation centres, indicating a role for Pfs230 in mediating enigmatic interactions between microgametes and erythrocytes (Eksi et al., 2006).

An interesting feature of anti-Pfs230 antibodies is that their TBA measured by SMFA was substantially enhanced by the presence of human complement (Read et al., 1994). Consistent with this observation was association between the presence of Pfs230-specific antibodies in malaria-exposed immune sera and the complement-mediated lysis of *P. falciparum* gametes (Healer et al., 1997). The temporal expression of Pfs230 in gametocytes, and its carriage over onto the surface of the gametes, which are formed within minutes of blood meal ingestion, permits interaction with the human complement in the blood meal. The fact that human complement is degraded by mosquito proteases within 3—5 h of blood meal ingestion explains the short-lived efficacy of anti-230 antibodies in the infected blood meal (Grotendorst and Carter, 1987; Margos et al., 2001).

Analyzing Pfs230 orthologous genes in eight *Plasmodium* spp. including *P. vivax*, Doi et al. (2011) discovered the structural conservation of the 14 cysteine motif/paired double domains that follow an interspecies-variable N-terminal pro-domain. Sequence polymorphisms were found among the 113 isolates examined; however, only a limited number of amino acid substitutions were found in a subdomain containing the first four cysteine motifs. This subdomain (Pvs230$_{236-943}$) is capable of eliciting TBA against *P. vivax* parasites from malaria patients in a membrane feeding assay, indicating this conserved structure may form the foundation of a candidate *P. vivax* transmission-blocking vaccine (Doi et al., 2011; Tachibana et al., 2012). Interestingly the transmission-blocking activities conferred by the anti-Pvs230 immune sera were comparable in the presence or absence of human complement (Tachibana et al., 2012), indicating a mechanism other than complement fixing by these anti-Pvs230 immune sera. Similarly complement enhanced only a fraction of malaria-exposed immune sera in suppressing *P. vivax* oocyst development in mosquitoes (Arevalo-Herrera et al., 2011; Mendis et al., 1987), though it is unclear whether the enhancement by complement was associated with Pvs230.

P48/45 belongs to the same cysteine-rich structural family as P230 (Gerloff et al., 2005). The full length Pfs48/45 protein has 448 amino acids, consists of a signal sequence, three cysteine motifs organized as one and half double domains and a putative glycosylphosphatidylinositol (GPI) anchor (Kocken et al., 1993). Pvs48/45 shares 56% sequence homology with Pfs48/45, and has similar domain structures based on the position of cysteine residues and predicted secondary structures (Woo et al., 2013). Similar to Pfs230, deletion of Pfs48/45 by gene knockout did not affect the development of gametes. However, the knockout parasites produced a significantly lower number of oocysts when fed to mosquitoes (van Dijk et al., 2001). Using knockouts of the Pfs48/45 orthologue in *P. berghei*, a rodent malaria model, Van Dijk et al. demonstrated a role for P48/45 protein prior to fertilization: without P48/45 protein the male gametes could not fertilize female gametes, whereas the females could still be fertilized by wild type males.

Following P48/45-mediated attachment of the male and female gametes, fusion of the membranes to form a zygote is mediated by HAP2/GCS1, a male-specific malarial protein originally identified in plants (Liu et al., 2008; Mori et al., 2010). Antibodies to a fragment of HAP2 from *P. berghei* expressed in *Escherichia coli* (HAP2$_{355-699}$) reduced in vitro formation of ookinetes by 81%, and, in the SMFA similarly reduced oocyst burden within the mosquito (Blagborough and Sinden, 2009). Subsequently using HAP2$_{194-604}$ of the *P. falciparum* orthologue expressed in the wheat germ cell-free system as immunogen, Miura et al. (2013b) demonstrated a significant reduction (97%) in infection intensity in mosquitoes.

Pfs25 and Pfs28 are paralogous proteins, most abundantly expressed postactivation on the surface of the developing zygote and ookinete. They possess a signal sequence, four epidermal growth factor (EGF)-like domains, followed by a C-terminal GPI anchor (Duffy and Kaslow, 1997; Kaslow et al., 1988). They are also cysteine rich and highly constrained by up to 11 disulfide bonds. Sequence analyses of Pvs25 and Pvs28, the orthologous proteins in *P. vivax*, revealed a similar structural organization though only 45% and 36% amino acid identity to Pfs25 and Pfs28, respectively (Tsuboi et al., 1998). Resolved crystal structures showed that Pvs25 formed tilelike trapezoid prisms, with four EGF domains lying in the same plane (Saxena et al., 2006), similar 3D structures were modelled for Pfs25 and Pbs28 (Sharma, 2008; Sharma et al., 2009). Using genetic knockout methods in the *P. berghei* model, Tomas et al. demonstrated that single deletion of either Pbs25 or Pbs21, orthologues of Pfs25 and Pfs28, respectively, had little effect on parasite development. However, deletion of both Pbs25 and Pbs21 drastically inhibited the formation of ookinetes, and reduced parasite capacity to penetrate the midgut wall and to form oocysts (Tomas et al., 2001). Surface expression of P25 protein is first detected on macrogametes, about 2 h after activation and prior to fertilization, though protein synthesis has been detected in cultured gametocytes (Vermeulen et al., 1986). Whether this expression results from artefactual derepression of the DOZI/CITH translation-inhibition, during gametocyte preparation has not been satisfactorily resolved. P28 proteins are synthesized slightly later than P25 and in *P. berghei* are detectable on unfertilized macrogametes (Sinden et al., 1987), and in *P. vivax* and *P. falciparum* are detectable on cell surface after zygotes elongate to become a 'retort' (Hisaeda et al., 2000; Vermeulen et al., 1986).

2.1.2 Sequence polymorphism of parasite targets in field isolates

The temporal expression of the 6-cys family of proteins during gametocyte development in human hosts (Vermeulen et al., 1985a) is consistent with the observation that malaria-exposed populations develop antibodies against

Pfs230 and Pfs48/45 after natural malaria infection (Bousema et al., 2010b; Graves et al., 1988b; Miura et al., 2013b; Ong et al., 1990; Ouedraogo et al., 2011; Premawansa et al., 1994; Riley et al., 1994). Consistent with these observations is the significant sequence polymorphism in Pfs230, Pvs230, Pfs48/45 and Pvs48/45 in field isolates, presumably as a result of immune evasion mechanisms by parasites under host immune pressure (Niederwieser et al., 2001; Williamson and Kaslow, 1993; Woo et al., 2013) despite the presence of functional and/or structural constraints that might have limited the levels of polymorphism (Doi et al., 2011). Analyzing sequences of a 76 kDa segment, encompassing the mature N-terminal of Pfs230 including the first 2 cysteine motifs, from 13 parasite strains, revealed 12 substitutions in 10 amino acid positions, all reside in cysteine motifs (Tachibana et al., 2011). Sixteen alleles in 40 amino acid positions in Pvs230 sequences were found in 112 isolates from worldwide locations, where polymorphisms were spatially segregated (Doi et al., 2011). Similarly, 24 alleles in 25 amino acid positions in Pfs48/45 sequences were found in 44 isolates from Kenya, India, Thailand and Venezuela. The isolates from Kenya were more polymorphic than those from other regions, indicating presence of greater positive selection pressure (Escalante et al., 2002).

Some reports have described the presence of P25-reactive antibodies in malaria-exposed populations (Kim et al., 2011a; Premawansa et al., 1994; Riley et al., 1994). The P25 proteins precipitated by these natural immune sera were recognized by a P25-specific monoclonal antibody, suggesting motifs found in P25 can be exposed to the human host (Riley et al., 1994). However, in contrast to relatively more polymorphic P230 and P48/45, only two single amino acid replacements (G131A, V144A) were found in the Pfs25 sequence, and a single amino acid replacement (K72R) was found in the Pfs28 sequence among isolates from various geographic regions (Da et al., 2013; Hafalla et al., 1997; Kaslow et al., 1989; Richards et al., 2006; Shi et al., 1992). Polymorphism of Pvs25 was also limited to 8 amino acid positions, but polymorphism of Pvs28 was found in up to 13 amino acid positions (Gonzalez-Ceron et al., 2010; Sattabongkot et al., 2003; Tsuboi et al., 2004; Zakeri et al., 2009).

2.2 Ookinete-secreted proteins

Upon maturation in the midgut, an ookinete in the blood meal must first penetrate the enclosing PM, interact with ligands on midgut microvilli, invade the midgut epithelium and emerge on the basal side of the cell and come to rest under the basal lamina where it forms an oocyst. Since these essential processes are totally dependent upon the activities of proteins secreted from ookinete organelles (notably the micronemes), these proteins became interesting targets for TBV development.

The PM is made of chitins, proteins and proteoglycans. The discovery of ookinete-secreted chitinases (Huber et al., 1991; Vinetz et al., 2000) prompted efforts to evaluate their role and function as a vaccine target. Disruption of expression of *Pfcht1* and *Pbcht1*, paralogous chitinase genes in *P. falciparum* and *P. berghei* parasites, respectively, resulted in marked reduction in oocyst formation in mosquitoes (Dessens et al., 2001; Tsai et al., 2001). A monoclonal antibody 1C3 against recombinant PfCHT1 recognized PgCHT2, an orthologous gene product in *P. gallinaceum*, and a recombinant 1C3 V_H and V_L assembled as a single chain was shown to reduce oocyst formation in mosquitoes fed on either *P. falciparum* gametocyte culture or on *P. gallinaceum*-infected birds (Li et al., 2005). Subsequent efforts were made to produce recombinant chitinases as vaccine candidates. Recombinant PfCHT1 and PvCHT1, the paralogous gene product in *P. vivax*, were produced in a wheat germ cell-free expression system (Takeo et al., 2009), the recombinant proteins displayed chitinase activity with differential substrate specificities. Antibodies raised against recombinant PvCHT1 recognized mosquito midgut stage parasites, whilst the enzyme was sensitive to the small molecule inhibitor allosamidin (Takeo et al., 2009), it is still unclear whether antibodies to the protein will elicit effective transmission-blocking immunity.

Several ookinete micronemal proteins were discovered by virtue of their predicted domain structure, such as the von Willebrand factor-A domain, known to be involved in cell–cell or cell–matrix interaction. Possible involvement of von Willebrand factor-A domain-related protein (WARP) in interaction with mosquito ligands was hypothesized, and PbWARP was first identified by fishing for coding sequences of the functional domain in *P. berghei* (Yuda et al., 2001). Subsequently WARP orthologues were cloned in other species including *P. falciparum* and *P. vivax* (Gholizadeh et al., 2010; Li et al., 2004). The protein is expressed in the lumen of ookinete micronemes, antibodies raised against PfWARP significantly reduced infection prevalence in both *Anopheles stephensi* mosquitoes fed on *P. falciparum*-infected RBC cultures, and in *A. aegypti* mosquitoes fed on *P. gallinaceum*-infected birds (Li et al., 2004).

The circumsporozoite- and thrombospondin-related adhesive protein (TRAP)-related protein (CTRP), features six von Willebrand factor-A like domains that are essential to ookinete gliding motility (Ramakrishnan et al., 2011) in addition to seven thrombospondin type 1 domains putatively involved in interaction with the mosquito ligands. CTRP is secreted onto the ookinete surface, and was shown to be essential to ookinete motility by gene disruption studies (Dessens et al., 1999), antibodies to PgCTRP significantly reduced infection prevalence in *A. aegypti* mosquitoes (Li et al., 2004).

Plasmodium perforin-like proteins (PPLPs) were identified by screening a cDNA library for putative membrane attack complex/perforin-like domain, a domain well characterized in mammalian pore-forming proteins and perforin (Kaiser et al., 2004). One of these proteins, membrane-attack, ookinete protein (MAOP) is located in the micronemes of ookinetes (Kadota et al., 2004). Disruption of MAOP gene expression in *P. berghei* parasites completely abolished ookinete traversal through the midgut epithelium. It is not yet known whether antibodies to MAOP have similar activity, but clearly this deserves further evaluation.

SOAP (secreted ookinete adhesive protein) and CelTOS (cell-traversal protein for ookinetes and sporozoites) are two additional ookinete microneme proteins, identified by screening the expression-tagged cDNA libraries for stage-specific expression. Disrupting expression of SOAP and CelTOS genes markedly impaired ookinete ability to penetrate the midgut wall, indicating their critical roles in parasite's differentiation from ookinetes to oocysts (Dessens et al., 2003; Kariu et al., 2006). Despite clear elucidation of their roles in ookinete and sporozoite development, and the ability of anti-CelTOS immunity to reduce sporozoite infectivity (Bergmann-Leitner et al., 2010), convincing evidence for TBA is lacking. There could be two explanations: (1) these proteins are stored in ookinete organelles and are released only when the ookinete makes intimate contact with target ligands. The resultant brief and late exposure to the antibodies might have made them ineffective targets; (2) the functional roles played by these proteins may be multiply redundant, such that blocking function of one target protein is insufficient to block transmission. This also indicated that gene disruption, whilst being a powerful tool in elucidating biological functions of the target protein, may not be an accurate predictor for an effective vaccine.

2.3 Mosquito components

Penetration of the midgut wall by ookinetes is a prerequisite to oocyst formation and subsequent oocyst development, and requires participation of (1) midgut structural ligands in cell recognition, attachment, gliding, cell entry, traversal and egress by the ookinetes, and (2) immune ligands regulating vector responses to infection. Vaccines targeted to mosquito midgut components are attractive candidates because the resultant reduction in vector competence may simultaneously block the transmission of multiple malarial species. In early proof-of-concept studies, presence of antibodies against midguts of *Anopheles farauti* resulted in a significant reduction in infection prevalence, with 3.4—51% mosquitoes infected when they fed on midgut-immunized and *P. berghei*-infected mice, compared to 60—80%, fed on the control mice (Ramasamy and Ramasamy, 1990). Using the SMFA a similar small, but significant reduction in infection prevelance was also observed by feeding *A. stephensi* mosquitoes on midgut microvilli-immunized and *P. berghei*-infected mice (Almeida and Billingsley, 2002), or by feeding *Anopheles tessellatus* mosquitoes with *P. vivax*-infected blood mixed with rabbit antisera or purified IgG raised against midguts of *A. tessellatus* (Srikrishnaraj et al., 1995). A stronger blocking of transmission of *P. falciparum* and *P. vivax* to *A. stephensi* and *Anopheles gambiae* mosquitoes was achieved by monoclonal antibodies raised against *A. gambiae* midguts. These mAbs recognized glycoproteins in *A. gambiae* midgut lysate, though their identities were not known (Lal et al., 2001).

Efforts had been made to identify mosquito components that may block malaria transmission. Lectins, such as Jacalin, known to bind to glycoproteins, were found to reduce ookinete attachment to the mosquito midgut by masking glycan ligands (Zieler et al., 2000). Using Jacalin-affinity chromatography followed by tandem mass spectrometry, Dinglasan et al. identified *A. gambiae* aminopeptidase (AgAPN1), a 125 kDa glycoprotein with a signal sequence, an insect gluzincin aminopeptidase motif and a GPI anchor sequence (Dinglasan et al., 2007). The N-terminal segment of 135 amino acids was expressed in *E. coli* and the polyclonal antibodies against recombinant product rAnAPN1$_{60-195}$ imparted parasite transmission blockade in mosquitoes fed on *P. berghei*-challenged mice, on *P. falciparum* gametocyte culture (Dinglasan et al., 2007), as well as on *P. falciparum*- or *P. vivax*-infected blood from patients in malaria endemic area (Armistead et al., 2014).

CPBAg1, a carboxypeptidase B has been identified by differential display screening of *A. gambiae* midgut transcripts (Bonnet et al., 2001). CPBAg1 expression is up-regulated and the carboxypeptidase activity in midguts increases when mosquitoes are infected with *Plasmodium*. Antibodies against a recombinant CPBAg1 reduced mosquito infection by gametocytes from *P. falciparum* carriers in membrane feeding assays. Reduction in infection prevalence and intensity was also observed in mosquitoes fed on

P. berghei-infected mice immunized with recombinant CPBAg1 (Lavazec et al., 2007). These data indicate CPBAg1 might also serve as a TBV target. However, the fact that the protein shares high similarity (48—54%) and identity (28—32%) with its mammalian homologue may make it a less attractive target for vaccine development (Lavazec et al., 2005).

Another proposed candidate target was a midgut trypsin involved in the activation of malarial chitinase (Shahabuddin et al., 1996). Indeed trypsin inhibitors and antitrypsin antibodies reduced infection to mosquitoes (Shahabuddin et al., 1995, 1996). However, like CPBAg1, the candidacy of trypsin as a vaccine target ended when it was discovered that the amino acid sequence of the active sites in trypsins from various species, including that of *Homo sapiens*, are highly conserved (Rypniewski et al., 1994).

A salivary gland protein of 100 kDa, saglin, was identified as a putative target based on recognition by a mAb that caused a 70% reduction in sporozoite intensity in mosquito salivary glands (Okulate et al., 2007). As a target ligand of SM1, a twelve-amino-acid peptide that resembles TRAP protein, saglin unsurprisingly also binds to TRAP with high specificity (Ghosh et al., 2009). Injection anti-saglin antibodies into the mosquito haemocoel, or reducing saglin expression by RNA interference resulted in significant reduction of sporozoite number in salivary glands. Thus saglin may play a crucial role in sporozoite invasion of the salivary glands through interaction with TRAP, and has been considered by some to be a potential target for TBV. It would be challenging, however, to deliver saglin-specific antibodies to haemocoel of the insect.

Following the recent analyses of the modulation of *Plasmodium* infectivity to the mosquito by the vectors' immune system (Molina-Cruz and Barillas-Mury, 2014), it has been suggested that the insect immune response could be up-regulated by antibodies delivered in the blood meal. Proof of concept has been demonstrated; antibodies against the *A. gambiae* serine protease inhibitor (serpin)-2 reduced *P. berghei* oocyst numbers by 54% (Williams et al., 2013), but this avenue has not yet been pursued further.

3. VACCINE DEVELOPMENT EFFORTS AND STATUS

To date, only a handful of candidates have been extensively evaluated for their feasibility as TBV targets. These include the parasite surface proteins P230, P48/45, P25 and P28, and the mosquito target AnAPN1. The 'Rainbow Tables' compiled and updated by the World Health Organization

(Schwartz et al., 2012), which provide a snapshot of malaria vaccine development pipeline, suggest the most advanced and somewhat surprisingly the only candidates, in the TBV pipeline are the recombinant Pfs25-based vaccines.

3.1 Challenge in production of recombinant vaccines

Despite the early proof-of-concept evidence that antibodies induced by P230, P48/45, P25 and P28 confer transmission-blocking activities, the unique cysteine-rich structural features of these protein imposed significant constraints on the choice of expression systems in which recombinant products might achieve native conformation.

Escherichia coli is a widely used expression vector capable of producing large quantities of recombinant peptides within inclusion bodies. However, the post recovery processing requires protein denaturation and refolding, where preservation of native conformation can be a challenge. Indeed, early efforts in producing recombinant Pfs25 in *E. coli* resulted in an improperly folded recombinant product not recognized by transmission-blocking mAbs, and antibodies induced by this recombinant Pfs25 failed to block transmission (Kaslow et al., 1992). It was only after redesigning and reconstructing a codon harmonized artificial gene encoding Pfs25 did the recombinant Pfs25 produced in *E. coli* successfully refold to retain native conformation. The mouse IgGs elicited by this product conferred 100% transmission-blocking activities in SMFA (Kumar et al., 2014).

Saccharomyces cerevisiae has also been widely used as an expression platform. The early data from two recombinant Pfs25 proteins produced in *S. cerevisiae* indicate that the yeast expression system might be able to replicate, at least partially, the conformation of the native protein. Pfs25—B_{22-188}, a recombinant Pfs25 with four EGF domains but lacking both signal sequence and GPI anchor, was secreted to the *S. cerevisiae* culture supernatant as a soluble protein. Pfs25—B was recognized by several conformation-dependent mAbs and elicited transmission-blocking antibodies in mice and monkeys (Barr et al., 1991; Kaslow et al., 1994). TBV25H, a hexa-histidine-tagged Pfs25, a full-length protein lacking both signal and anchor sequences, was also secreted into the *S. cerevisiae* culture supernatant as a soluble protein (Kaslow and Shiloach, 1994). TBV25H formulated with aluminum hydroxide was tested in two phase 1 trials in humans. The first trial was terminated upon observation of atypical hypersensitivity reactions, likely due to free antigens dissociated from alum (Gozar et al., 2001; Kaslow, 2002). However, in the second trial in which a viral vector prime (NYVAC-Pf7) was boosted with

TBV25H/alum, some volunteers developed transmission-reducing antibodies (Kaslow, 2002). Pvs25 has since been produced in *S. cerevisiae* as secreted protein (ScPvs25). Induction of transmission-reducing activities in rhesus monkeys and in humans when ScPvs25 was formulated with Alhydrogel® and Montanide ISA 720 (Malkin et al., 2005; Saul et al., 2007), indicated the presence of properly folded recombinant ScPvs25.

Despite having been tested in multiple human trials, recombinant P25 produced in *S. cerevisiae* remained heterogeneous in conformation, among which a homogeneous form (A-form) recognizable by a conformation-dependent mAb represented as little as 10% of the production total (Miles et al., 2002; Zou et al., 2003). The *Pichia pastoris* expression system overcame this shortcoming by producing >60% A-form Pfs25, this production yield was further improved by overexpressing a *P. pastoris* protein disulfide isomerase in the expressing cells (Tsai et al., 2006; Zou et al., 2003). This form of PpPfs25 was tested formulated with Montanide ISA 51 in a phase 1 trial. Although the trial was terminated when the ScPvs25 arm of the trial had unexpected systemic reactions, one of five volunteers who completed the two scheduled vaccinations of 5 μg PpPfs25H/Montanide ISA51 developed potent transmission-reducing activities (Wu et al., 2008).

Recently Pfs25 was successfully expressed in a plant-based launch vector system as a fusion to the lichenase carrier molecule (Pfs25—LiKM) (Farrance et al., 2011a) and as a fusion to the *Alfalfa* mosaic virus coat protein (Pfs25—CP) (Jones et al., 2013). In each case, the recombinant products were purified to a high level of homogeneity. Pfs25—CP self-assembled as viral-like particles (VLPs) highly consistent in size. Mice immunized with one or two doses of Pfs25—CP VLPs developed serum antibodies with complete transmission-blocking activities throughout the 6 month follow-up period. A phase 1 trial with Pfs25—CP VLPs is underway (ClinicalTrials.gov Identifier: NCT02013687).

The lasting functional antibody titres induced by Pfs25—CP VLPs may rekindle interests to re-evaluate other expression systems such as baculovirus. In fact, baculovirus was one of the expression platforms tested in early years, for the production of a full-length GPI-anchored recombinant $Pbs21_{1-218}$, (the *P. berghei* P28 protein) on the host cell surface. The product elicited strong transmission-reducing activities in mice, but it was difficult to purify (Blanco et al., 1999). To facilitate the purification process, the C-terminal hydrophobic GPI anchor signal was removed to generate a secreted version, $Pbs21_{1-188}$. However, $Pbs21_{1-188}$ was a weaker immunogen and poorly recognized by a conformational mAb, indicating that the GPI anchor sequence might both facilitate the protein folding to native conformation, and act as an adjuvant. Now that the baculovirus expression vector systems (BEVS) are much improved (Fernandes et al., 2013), and licensure of Cervarix™ had set precedent for regulatory approval of products made by this platform, it is worthwhile to re-evaluate the BEVS for recombinant production of malaria vaccine targets. Encouragingly, a recombinant baculovirus expressing Pfs25 on its surface envelop induced strong transmission-reducing activities as demonstrated by passive IgG transfer to, as well by active immunization of, mice challenged with a transgenic *P. berghei* parasite expressing Pfs25 (Mlambo et al., 2010). As a recent advance in baculovirus expression platform, baculovirus dual expression system (BDES) has been developed based on the baculovirus *Autographa californica nucleopolyhedrosis virus*, which possesses strong adjuvant properties that can activate dendritic cell-mediated innate immunity. Rabbit IgGs elicited by BDES—Pvs25 conferred strong reduction (98%) in *P. vivax* infection intensity in a membrane feeding assay in which *Anopheles dirus* mosquitoes were offered to feed on a blood meal composed of the test IgGs and blood from Thai gametocyte carriers (Blagborough et al., 2010). The BDES-Pvs25 was also evaluated in an in vivo assay in which the BDES-Pvs25-immunized mice were inoculated with a transgenic *P. berghei* expressing *P. vivax* Pvs25, followed by direct mosquito feed on these mice, and assessment of infections in these mosquitoes. In one study, intranasal (IN) and intramuscular (IM) immunization of BDES-Pvs25 achieved a 92% and an 88% reduction of infection intensity, and an 84% and a 76% reduction in infection prevalence, with IN and IM deliveries, respectively (Blagborough et al., 2010). In another study, 56% and 60% reduction in infection intensity and prevalence, respectively, were achieved (Mizutani et al., 2014).

Production of recombinant Pfs230 was hampered by its size of approximately 310 kDa, in addition to the 14 cysteine motifs constrained by disulfide bonds. In the search for sub-domains with transmission-blocking activities, Williamson et al. (1995) evaluated four segments of Pfs230 expressed in *E. coli* as fusions to a maltose-binding protein (MBP). The 76 kDa 'segment C', encompassing amino acids 443−1132 (Pfs230C$_{443-1132}$) with the EEVG repeat and the first two cysteine motifs, was the only segment that conferred transmission-reducing activities. To further dissect the Pfs230C$_{443-1132}$ segment, Tachibana et al. (2011) produced recombinant segments C$_{443-1132}$, C0$_{443-588}$, C1$_{443-715}$ and C2$_{443-915}$ in a wheat germ cell-free protein expression system. All these segments, including the C2$_{443-588}$ which was truncated just before the start of the first cysteine motif, induced rabbit antibodies that confer reduction in infection intensity ranges from 66% to 99% in SMFA (Tachibana et al., 2011). The activities were, as expected, enhanced by the presence of complement, reminiscent of the enhancement of transmission-blocking and -reducing activities conferred by monoclonal antibodies (Read et al., 1994; Roeffen et al., 1995). The finding is particularly interesting for the development of vaccines targeting to Pfs230: the functional pro-domain $C0_{443-588}$ resides outside of polymorphic region, and is therefore likely to induce strain-transcending immunity. Based on this finding, a process was developed to produce a $Pfs230_{444-730}$ in a plant-based launch vector system. The recombinant segment induced strong transmission-blocking activities in rabbits in a complement-dependent manner (Farrance et al., 2011b). A scaled production process to produce a shorter Pfs230 in *P. pastoris* was also developed, and this recombinant product also induced strong transmission-blocking activities (MacDonald, Narum, and Wu, unpublished). This Pfs230 subdomain is being evaluated in a phase 1 trial (ClinicalTrials.gov Identifier: NCT02334462).

Production of properly folded recombinant P48/45 that can elicit transmission-blocking immunity proved to be more challenging. Early attempts were made with a variety of expression platforms including *E. coli*, *S. cerevisiae* and *P. pastoris*. The recombinant products were either not detected (in *S. cerevisae*) or mis-folded (in *E. coli* and *P. pastoris*) and did not induce transmission-blocking immunity (Milek et al., 1998b; Milek et al., 2000). A recombinant vaccinia viral vector containing the Pfs48/45 gene was also used to test expression in mammalian cells. Despite the presence of signal sequences and GPI anchor, the recombinant protein was trapped inside Hela cells, and infection of animals with this recombinant virus failed to elicit transmission-blocking immunity (Milek et al., 1998a). More recently encouraging results were achieved using the *E. coli* expression platform, probably at the expense of production yield, when formation of inclusion bodies was avoided. When co-expressed with *E. coli* periplasmic folding catalysts, a recombinant $Pfs48/45_{159-428}$ fused to MBP, encompassing the C-terminal two cysteine motifs of Pfs48/45, is secreted to the periplasmic space of the *E. coli* cells as a soluble protein (Outchkourov et al., 2008). The purified product elicited strong transmission-blocking activities following as many as five vaccine doses. The same $Pfs48/45_{159-428}$ was also fused to an asexual stage malaria antigen glutamate-rich protein, and expressed in a Gram-positive bacterium *Lactococcus lactis*. The Pfs48/45 moiety in the chimeric product seemed to be correctly folded and contain native, transmission-blocking epitopes, as evidenced by induction of

TRA in rats, even when the recombinant chimera was formulated with aluminum adjuvant (Theisen et al., 2014).

A different approach was taken to 'harmonize' Pfs48/45 codons to those used in *E. coli*, so that the rate of recombinant protein synthesis could be coupled with proper folding. In addition, by treating the cell lysate with ionic detergent, the expressed recombinant Pfs48/45 could be kept in a soluble state (Chowdhury et al., 2009). The full-length recombinant Pfs48/45 was recognized by conformation-dependent transmission-blocking mAbs, and elicited strong transmission-blocking activities in mice and in baboons.

Contrary to the cysteine-rich parasite surface proteins, AnAPN1 lacks cysteine in its amino acid sequence and has a relatively simple secondary structure (Armistead et al., 2014). As a consequence *E. coli* was chosen for production of recombinant $AnAPN1_{60-195}$ as inclusion bodies for the high yield and the facile downstream processing. The solubilized and refolded $rAnAPN1_{60-195}$ was immunogenic and retained properly folded transmission-reducing epitopes (Mathias et al., 2012). Despite 29% sequence identity shared between the $rAnAPN1_{60-195}$ and a putative human orthologue alanyl (membrane) aminopeptidase, there appears to be no cross-reactivity to human tissues by antibodies against rAnAPN1. However, since the in vitro immunohistochemistry cannot assess in vivo responses including autoimmunity after immunization, caution may need to be exercised when advancing the rANAPN1 vaccine candidate to human trials.

3.2 Challenges in vaccine formulation

Recombinant subunit vaccines in general require enhancement by adjuvants to achieve adequate immunity. Most of the initial proof-of-concept in vivo immunogenicity studies were conducted using recombinant proteins formulated with Freund's adjuvant, known to be a potent enhancer *for* in vivo immune responses; however, it is unsuited for use in humans. Because the benefit offered by TBVs is delayed for vaccinees, and a TBV will require higher compliance in vaccination coverage in order to achieve herd immunity (see below), a conservative approach has been taken in selecting adjuvants for TBV formulation: the adjuvant has to possess a broad safety profile in humans in addition to immune enhancement characteristics. This approach, however debatable, has excluded multiple experimental immune enhancers from use with TBV candidates now under clinical development.

Aluminium salts had been the only adjuvants used in licensed human vaccines until MF59, a squalene-based oil-in-water adjuvant, was approved in a flu vaccine in Europe in 2009. Like most recombinant protein-based malaria vaccine candidates, Alhydrogel® was the adjuvant of choice for clinical development of ScPvs25, the first TBV candidate that completed a phase 1 trial. The vaccine was well tolerated and TRA was observed by membrane feeding assays using *P. vivax*-infected blood from malaria patients in Thailand (Malkin et al., 2005). However, the activity was, at the time, considered insufficient to block transmission by mosquitoes and thus inadequate to underpin an effective vaccine. The need for more effective adjuvants was made clear by the generally inadequate immune responses to synthetic and recombinant malarial antigens seen with the use of aluminium-based adjuvants. Adding CpG enhances immunogenicity (Qian et al., 2008), but its use in humans was shadowed by a case of Wegener disease in a healthy volunteer included in a trial for hepatitis B vaccine with CpG-oligodeoxynucleotide (ODN), which led to a premature stop, in healthy volunteers, of all clinical trials using CpG-ODN in vaccines (DeFrancesco, 2008).

Montanide™ ISA720 (ISA720), and Montanide™ ISA51 (ISA51) are two potent investigational adjuvants used in many cancer vaccine trials (Aucouturier et al., 2002). Composed of squalene (ISA720) or light mineral oil (ISA51) and highly purified mannide mono-oleate emulsifier, both adjuvants form water-in-oil emulsions. ScPvs25/ISA 720 emulsion was shown to induce transmission-blocking activities in rhesus monkeys (Saul et al., 2007). ISA720 was the adjuvant in several clinical trials testing recombinant protein-based malaria vaccine candidates (Herrera et al., 2011; Hu et al., 2008; Lawrence et al., 1997; McCarthy et al., 2011; Oliveira et al., 2005; Saul et al., 1999, 2005). In these studies, the formulations were generally well tolerated, except for local reactogenicity thought to be related to formation of depots at the site of injection. However, the rate of the local reactogenicity in one of these studies was deemed unacceptable and the clinical trial was halted prematurely (McCarthy et al., 2011). Premature termination of another clinical phase 1 trial testing ScPvs25 and PpPfs25 formulated with ISA51, occurred where 2 cases of erythema nodosum in the Pvs25 arm were graded as 'probably related to vaccination' (Wu et al., 2008).

In perspective, new adjuvants under development will become of key interest for testing their adjuvanticity with TBV candidates. It will be particularly interesting to test MF59 (Novartis) that has been licensed for use in a flu vaccine, and AS01 (GlaxoSmithKlie's Adjuvant System-01) that has been tested extensively and is well tolerated in RTS,S, a partially protective malaria vaccine that is likely to be licensed.

3.3 Alternative approaches to enhance immunogenicity of candidate vaccines

In light of limited choice of adjuvants with acceptable safety profiles in humans, an alternative has been to modify the antigens themselves to make them more immunogenic while retaining their functional epitopes. Chemical conjugation of polysaccharides to protein carriers has turned the T-cell independent polysaccharide vaccines to T-cell dependent vaccines effective in young children (Kelly et al., 2004). These conjugate vaccines were shown to have the ability to elicit antibodies with high avidity and establish memory responses. Conjugating the beta subunit of human chorionic gonadotropin and an alpha subunit of ovine luteinizing hormone to protein carriers was also shown to improve antibody avidity (Talwar et al., 1994). The protein carriers used in licensed vaccines include: cross-reactive material of diphtheria detoxified toxin bearing the amino acid substitution (CRM_{197}); tetanus toxoid (TT), and meningococcal outer membrane protein complex (OMPC) (Pichichero, 2013). Another carrier, recombinant *Pseudomonas aeruginosa* detoxified toxin exoprotein A with amino acid 552 and 553 substitutions (rEPA), is not yet a component of any licensed vaccine, but has been tested extensively in humans including infants (Lin et al., 2001; Passwell et al., 2003, 2010; Thiem et al., 2011).

Encouraging results have been obtained by conjugating PpPfs25 to OMPC: the conjugation significantly enhanced the antibody titres in mice, rabbits and rhesus monkeys. Chemical coupling did not seem to affect functional epitopes, as the correlation between the conjugate-induced antibody titres and their transmission-reducing activities was similar to that induced by PpPfs25 formulated with ISA51. More interestingly, the antibody response in rhesus monkeys induced by the conjugate sustained for 2 years, and recalled by a booster dose of PpPfs25 formulated with an aluminium adjuvant (Figure 2) (Wu and Craig, 2006). Subsequently, conjugating recombinant P25 and P28 to rEPA was also shown to enhance antibody responses in mice, which was further augmented by adding CpG into the formulation (Qian et al., 2007, 2008, 2009). A production process was developed to manufacture PpPfs25–rEPA conjugate in cGMP compliance (Shimp et al., 2013), and the Alhydrogel formulation of the Pfs25–EPA conjugate has been tested in Phase 1 trials in USA and Mali (CllinicalTrials.gov Identifiers: NCT01434381 and NCT01867463).

Miyata et al. (2010) reported that chemical conjugation of nontoxic cholera toxin B subunit with recombinant Pvs25 protein increased the

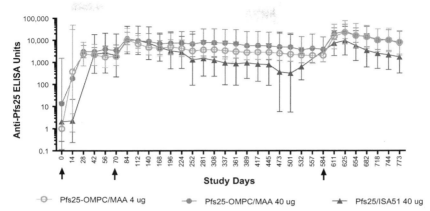

Figure 2 Recall response in rhesus monkeys immunized with Pfs25−OMPC *(Modified from Wu et al. (2006).)*. Rhesus monkeys were immunized with Pfs25−OPMC/MAA or Pfs25/ISA51 on D0 and D70, and a booster dose of Pfs25/MAA was given to all groups on D584 (as indicated by arrows).

transmission-reducing efficacy against *P. vivax*. They also developed a novel strategy to increase immune responses by creating genetic fusion proteins to target-specific antigen-presenting cells (APC). The fusion complex was composed of three physically linked molecular entities: (1) a vaccine antigen, (2) a multimeric alpha-helical coiled-coil core and (3) an APC-targeting ligand linked to the core via a flexible linker. Immunization of mice with the tricomponent complex fused to Pvs25 induced a robust antibody response and conferred substantial levels of *P. vivax* transmission blockade as evaluated by direct membrane feeding assay (DMFA) (Miyata et al., 2011).

In addition to the 'carrier' effect, epitope repetition in conjugates might be another mechanism of immune enhancement, similar to that of VLPs. Kubler-Kielb et al. (2007) demonstrated that polymerizing Pfs25 by chemical conjugation to itself improved Pfs25 immunogenicity and response longevity in mice. The multimeric Pfs25 core, in addition to its own merit as a target antigen, was also used as a carrier to enhance immunogenicity of the *P. falciparum* circumsporozite protein NANP repeat peptides (Kubler-Kielb et al., 2010), though the mechanism of the 'carrier' effect of the multimeric Pfs25 core is not known.

A second alternative strategy may be heterologous prime-boost using viral vectored products. This strategy has two obvious advantages: (1) viral vectors are more likely to express properly folded target antigen because they can co-opt the necessary chaperons in the host cell to assist target protein folding, as evidenced by Pfs25 protein expressed on the surface of HeLa cells by NYVAC-Pf7 (Tine et al., 1996); and (2) prime-boost regimens are known for their capacity for T-cell activation which might help to improve the quality and duration of the induced immunity (Hill et al., 2010; Nolz and Harty, 2011). Despite these advantages, the strategy was shadowed by concerns of potential cost, and to date only limited work has been published to evaluate this strategy on TBVs: work reported by Goodman et al. evaluated heterologous prime-boost with multiple viral vectors, including a chimpanzee adenovirus 63 (ChAd63), human adenovirus serotype 5 (AdHu5) and modified vaccinia virus Ankara (MVA)- constructed to incorporate a full length Pfs25 gene. Mice primed with ChAd63 or Adhu5 and boosted with MVA generated strong transmission-reducing activities, as shown in an in vivo assay where immunized mice were inoculated with a transgenic *P. berghei* expressing *P. falciparum* Pfs25, followed by assessment of mosquito infectivity after feeding on these mice (Goodman et al., 2011). Subsequent studies have compared the efficacy of this system to induce effective immunity against Pfs230, Pfs25 Pfs48/45 and AgAPN1 (Kapulu, Sinden et al., unpublished).

Biodegradable polymeric nano or microparticles may represent a promising new technology for improved immune responses (Gregory et al., 2013). Whereas the technology is quite far from application in clinical testing, the ability to sustain and control the release of the entrapped antigens and their delivery to APC raises hope for eliciting long-lasting immune responses. Immunizing mice with AnAPN1 entrapped in biodegradable microparticles prepared from polylactofate, Dinglasan et al. (2013) observed TRA, as measured by direct feeding of mosquitoes on immunized mice challenged with *P. berghei*, even 6 months after a single dose.

3.4 Multicomponent vaccines

Despite extensive data in multiple parasite—vaccinee combinations, demonstrating the ability of Pfs25 to induce transmission-reducing and blocking responses, the current formulation, PpPfs25—EPA/Alhydrogel, is unlikely to be an effective, standing alone TBV for use in man that could meet a currently envisaged Target Product Profile (TPP) (Alonso et al., 2011; Brooks et al., 2012). Whether the next candidate, Pfs25—VLP/Alhydrogel, can overcome the hurdle of immunogenicity and response-longevity awaits results in human trials.

Limited efficacy of the Pfs25 TBVs might be due to presence of a functionally redundant protein (Pfs28) in the parasite. Indeed, as demonstrated by single- or double-knockout of P25 and P28 genes in *P. berghei* parasites, the P25 and P28 proteins share partially redundant function in ookinete and oocyst development, and it is only the deletion of both P25 and P28 that resulted in significant impairment of oocyst development in mosquitoes (Tomas et al., 2001). Thus combining P25 with P28 or another target antigen as one TBV may overcome the current poor efficacy which might result from escapes from anti-single component immunity. It has been speculated that the combination may also decrease the proportion of poor responders to individual components (Saul and Fay, 2007). With progress made in production of functionally effective recombinant Pfs230 a phase 1 trial evaluating Pfs25—EPA/Alhydrogel® and Pfs230—EPA/Alhydrogel®, given separately and in combination, is in progress (ClinicalTrials.gov Identifier: NCT02334462). Using BDES (Section 3.1) viral constructs were made to deliver Pvs25 and PvCSP, fused with a liker sequence, to mice. A significant protection (P < 0.05) against challenge by transgenic *P. berghei* sporozoites expressing PvCSP on their surface was observed. Also observed was a significant reduction (P < 0.05) in both infection prevalence and intensity after mosquitoes fed on immunized mice infected with a transgenic *P. berghei* expressing Pvs25 (Mizutani et al., 2014).

4. PROTECTIVE CORRELATES AND SURROGATE ASSAYS FOR EFFICACY

How do we measure the impact of a transmission-blocking vaccine appropriately? The key property of a TBV is that it is designed to reduce the number of secondary cases of human infections caused by a case of malaria, a quantity referred to as the reproduction number, R. The ability of a control intervention to reduce R is termed the 'effect size' and this can be used to compare control interventions which act upon different sections of the life cycle and ultimately predict the public health benefit of a TBV candidate (Smith et al., 2007). It is not feasible to measure effect size or R directly in the laboratory in humans for a TBV in the same way it is typically done for pre-erythrocytic vaccines as it would require the infection of a large number of volunteers. To overcome this, scientists have instead used intermediate measures to quantify the ability of TBV candidates to reduce transmission (i.e. estimating the impact of reducing human—mosquito transmission instead of human—mosquito—human transmission). Here we

4.1 Antibody titre

By design, the vaccine-induced antibodies ingested by mosquitoes are the key effectors for TBVs aiming to block parasite development in mosquitoes, thus their titres can be quantified by antigen-specific ELISA. However, due to the polyclonal nature of the antibodies induced by recombinant protein-based vaccines, the ELISA titre alone maybe insufficient to quantify the levels of antibodies specific to a repertoire of functional epitopes yet to be identified. If a quantitative correlation can be established between the antibody titres and their biological function, e.g. TBA (e.g. reduction in the number of mosquitoes infected), ELISA could be an appropriate surrogate assay for high-throughput evaluation of a TBV (Miura et al., 2007). These results could be combined with mathematical models to predict the public health impact of a given candidate (Bousema et al., 2010a). It should be noted that, however, antibody titres should not be simply translated into a TBV efficacy as antibody titres typically vary substantially among vaccinated individuals (Saul and Fay, 2007), which will influence the population-level efficacy achieved (Saul, 2008).

4.2 Mosquito infection in laboratory feeding assays

The impact of transmission-blocking antibodies on parasite transmission has been assayed at various biological (surrogate) stages including the ookinete (Ramjanee et al., 2007), oocyst (Carter and Chen, 1976; Gwadz, 1976) and sporozoites in the salivary glands (Kim et al., 2011b; Lensen et al., 1992). Of these, perhaps the most commonly used is the SMFA, in which a gametocyte culture mixed with test antibodies or drugs is offered to a laboratory mosquito colony through an artificial membrane feeder (currently the wax membrane Parafilm® or its equivalent). The mosquitoes are maintained for 7–8 days before being dissected to determine whether oocysts have developed. Oocysts are enumerated on the assumption that all mosquitoes that develop oocysts are likely to go on to produce infectious salivary gland sporozoites, though there has been discussion as to whether the intensity or prevalence of the oocyst infection is the key criterion to measure (see below). This assay has been frequently described as a 'gold standard' assay to quantify the impact of antibodies on *P. falciparum* transmission to mosquitoes, and has been applied to evaluate nearly all *P. falciparum* TBV candidates from initial proof-of-concept animal studies to clinical trials.

A variant of this assay — the DMFA differs from SMFA in that the venous blood from gametocyte carriers is offered to laboratory-reared mosquitoes in the membrane feeder. The biological function of the test antibodies can be evaluated by replacing the autologous plasma of the gametocyte donor with the test antibodies. The DMFA is particularly valuable for evaluation of transmission-blocking activities against *P. vivax* for which infected blood from *P. vivax* patients is the only reliable source of gametocytes (Malkin et al., 2005; Ponsa et al., 2003).

Though the methodology for conducting the SMFA and DMFA is fairly well defined (Blagborough et al., 2013b) there is still considerable ambiguity over how the results of the assay are interpreted. Three possible endpoints have been used to describe the efficacy of a TBV to reduce human-to-mosquito transmission: (1) the reduction in the proportion of infected mosquitoes; (2) the reduction in the average number of oocysts in all mosquitoes and (3) the reduction in the average number of oocysts in infected mosquitoes. Each oocyst is capable of producing thousands of salivary gland sporozoites so it is assumed that all infected mosquitoes are equally infectious irrespective of how many oocysts they harbour (Stone et al., 2013). Were this assumption to be correct this would make oocysts prevalence, (1), the best measure for assessing TBV effectiveness epidemiologically speaking. Unfortunately the published literature clearly shows that the probability of the vertebrate host becoming infected is influenced by the number if sporozoites inoculated, thus oocyst number is a dependent variable. Oocyst intensity (either 2 or 3) has been shown to be a more sensitive metric in both SMFAs and DMFAs so is often used by laboratory scientists. The reason for this increased sensitivity is caused by the relationship between oocyst prevalence and intensity (Figure 3(a)) (Churcher et al., 2012; Medley et al., 1993). The number of oocysts in a mosquito is highly overdispersed (aggregated) with some mosquitoes having a high number of oocysts whilst others have relatively few. In mosquitoes that ingest a large number of parasites, a TBV may reduce the number of oocysts by a high percentage but fail to completely remove all oocysts. Using oocyst prevalence as a measure the efficacy in this mosquito would give a value of zero, whilst if oocyst intensity was used it could be close to 100%. This means that efficacy estimates based on oocyst prevalence are more dependent on the degree of parasite exposure (challenge intensity) This was shown using an anti-Pbs28 monoclonal antibody against *P. berghei* where repeatedly running SMFAs with the same reagent saw a reduction in prevalence efficacies with increasing parasite exposure whilst efficacy based on oocyst

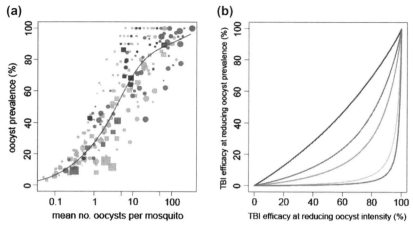

Figure 3 (a) The relationship between oocyst prevalence and intensity and (b) how this relationship will determine the effectiveness of transmission-blocking interventions *(Reproduced from Churcher et al. (2012).)*. In (a) point colour denotes parasite species, *Plasmodium falciparum* (blue (black in print versions)) or *Plasmodium berghei* (green (grey in print versions)). The shade of the colour indicates whether the experiment was a control (lighter) or included a candidate transmission-blocking intervention (darker). The size of the points signify the number of mosquitoes dissected whist the shape denotes the species of vector used, *Anopheles gambiae* (square) or *Anopheles stephensi* (circle). The black line illustrates the best fit model which is used to predict how transmission-blocking vaccine (TBV) efficacy against oocyst prevalence and intensity will vary according to the mean oocyst intensity in the control group (b), be it one (blue (black in print versions)), five (green (dark grey in print versions)), 10 (orange (light grey in print versions)), 50 (yellow (very light grey in print versions)) or 100 (red (grey in print versions)) oocysts.

intensity (in all mosquitoes) remained roughly constant (Churcher et al., 2012). In wild-caught mosquitoes, oocyst intensity lies at the lower end of the variable range (Beier et al., 1992; Billingsley et al., 1994; Gouagna et al., 2013). Thus we must be aware the efficacy reduction observed in SMFAs with high oocyst intensity in the control feeds, as a result of high parasite challenge, may underestimate the epidemiological impact of the TBV candidate in the field setting.

The SMFA and DMFA are complex and involve multiple biological systems all with inherent variability. Perhaps the most critical prerequisite in establishing quantitative relationship between antibody titres and the biological function is to minimize the assay variability and improve the reliability of estimated biological activity. By analyzing data from over 100 SMFAs testing sera from malaria-endemic populations among which many repeated measurements were obtained (van der Kolk et al., 2005), and data from repeated SMFAs testing a mouse anti-Pfs25 monoclonal antibody 4B7 at multiple concentrations (Miura et al., 2013a), the authors estimated the inter- and intra-assay variability. The inter-assay variability may be attributed to batch effects from parasites or mosquitoes used in individual experiments. The variability may be reduced by controlling the parasite exposure to mosquitoes by only accepting an experimental feed when oocyst intensity in the control feed reaches a certain number. Comparing samples within the same experimental feed and repeating the experiment multiple times improves the reliability of the data. Overdispersion in oocyst distribution in individual mosquitoes may explain, at least in part, the intra-assay variability when coupled with relatively few mosquitoes being dissected. Empirically van der Kolk et al. (2005) suggested that increasing the number of mosquitoes to be dissected and examined for infection, could reduce the intra-assay variability and improve the reliability of the estimated biological activity. However, the exact number of mosquitoes needed will vary according to the question under investigation and the definition of efficacy used. For example, precise estimates of changes in the reduction in the average number of oocysts in all mosquitoes (efficacy estimate 2 above) will require a large number of mosquitoes to be dissected as inclusion of the rare mosquitoes with very high oocyst intensity overly influences efficacy estimates. However, a smaller numbers could be dissected if approximate estimates would suffice or if efficacy was measured as a reduction in the number of infected mosquitoes (efficacy estimate 1 above). Given the variability in sample sizes used, all efficacy estimates should include confidence intervals to show the uncertainty around the point estimate. Using statistical methods such as generalized linear mixed-models will enable more precise estimates to be presented without the need to increase sample sizes (Churcher et al., 2012; Miura et al., 2013a).

The DMFA has a clear advantage over the SMFA in that the gametocytes, being derived from infected hosts, reflect the variability of the source population bringing more variables into the assay (Bousema et al., 2012; Da et al., 2013; Graves et al., 1988a; Malkin et al., 2005; Toure et al., 1998). This variation can provide important insights into the potential impact of an intervention in different field populations. The DMFA may also be conducted directly on whole blood of gametocyte carriers who have received TBV candidates. In such assays, transmission-blocking efficacy would be estimated in comparison with mosquito infection intensity or prevalence from DMFA conducted on volunteers receiving a control vaccine, though considerable care must be exercised in the experimental design to remove the variability in both parasite challenge, and confounding host factors. This important feature of the DMFA allows parallel analyses to link SMFA to another field-based feeding assay, e.g. the direct skin feed (DSF). In DSF assays, laboratory-reared mosquitoes are allowed to take a blood meal from the microvasculature of gametocyte carriers by direct skin probing (Bonnet et al., 2000; Toure et al., 1998). Obviously DSF would be closer to the natural conditions than the DMFA, but has lower community acceptance. The suitability and feasibility of DMFA and DSF for TBV evaluation were comprehensively reviewed by Bousema et al. (2012) where the authors analyzed data of hundreds of published and unpublished DMFA and DSF experiments conducted in Africa. They found that blood meals from the same carrier resulted in a higher proportion of infected mosquitoes by DSF compared to that by DMFA, indicating transmission by DSF is more efficient and DSF is a more sensitive assay. However, what is not clear is whether this difference in transmission will have any meaningful impact on the evaluation of different transmission-blocking interventions.

4.3 Laboratory-based population assay

Though immensely useful for understanding transmission biology and selecting good TBV candidates, it is essential we acknowledge that, at present, we do not know how the surrogate endpoints of all the above assays correlate with effect size and our ability to reduce the key parameter R. What we do know from animal model is that reductions in either gametocytes, ookinetes or oocysts numbers (intermediate parasite life stages) do not linearly correlate even with a reduction in sporozoite numbers in the salivary glands of mosquitoes (Sinden et al., 2007). If these negative density-dependent processes acting between female gametocyte number and ookinete; ookinete and oocyst, and oocyst and salivary gland sporozoite number occur in natural wild parasite—vector combinations then it is likely that assays measuring these intermediate life stages would overestimate of the effectiveness of a partially efficacious TBVs (Churcher et al., 2010).

In an effort to embrace all these unknowns in a transmission-blocking population assay, and directly estimate the effect size, Blagborough and colleagues developed a multigenerational population transmission assay using cloned *P. berghei* parasites transmitted between inbred lines of both mice and mosquitoes. In the first report of the assay, they describe how a drug (atovaquone) when used at a concentration that reduced oocyst intensity and prevalence in direct feeding assays by 57% and 32% respectively nonetheless eliminated the parasite from the population in just three transmission cycles, but only at low biting rates (Blagborough et al., 2013a). Thus successive (multicycle) use of an antimalarial transmission-blocking intervention can result in elimination of *Plasmodium* at low transmission levels. Significantly, estimates of the effect size in this first study were only 20%, substantially less than the impact predicted by the direct feeding assay. This suggests that there are frequency- or density-dependent process acting between oocyst development and mouse patency that reduce the value of the former in estimating the population level efficacy of a TBV (Churcher et al., 2010). Whilst care must be taken extrapolating these laboratory studies to natural parasite—vector combinations, it does suggest that feeding assays might overestimate the effectiveness of a TBV candidate in field settings (given the same exposure to infection).

4.3.1 TPP and the path forward
Defining the TPPs has been used as a strategic tool to guide product research and development in the pharmaceutical industry. With the aim of using a TBV as an in integral component of malaria elimination measures, an international consultation consisting of malaria researchers, policy makers and funding agencies reached a consensus on the primary endpoint to be used for vaccines interrupt malaria transmission. A TPP for TBVs presumed the need for >85% efficacy (reduction in mosquito infection prevalence) and a minimum protective duration of 2 years (Alonso et al., 2011; Brooks et al., 2012; Dinglasan et al., 2013). Given the current status of TBV development it is unsurprising that attention would be focused on the estimating the efficacy of the intervention (as measured by feeding assays). A key aspect cannot be overlooked is the duration of protection, because the persistence of transmission reduction and the rate and shape of its decline might have a greater impact on the public health benefit and cost-effectiveness of a candidate than initial efficacy. As noted in the population assay above and mathematical modelling work (Smith et al., 2011; Wenger and Eckhoff, 2013) even partially effective TBVs are likely to result in elimination in areas of low endemicity (either in sites where the pre-control biting rate was low, or where potential control interventions have been deployed effectively), indicating a need to revise the current TPP to reflect epidemiological settings in which the TBV product should be effective.

In addition to advancing existing TBV candidates along the development pathway and demonstrating their protective efficacies in various surrogate assays, new target identification and characterization is critical to feed the upstream steps of the development pipeline. Challenges remain in our pathway forward from the design of pivotal clinical trials (Delrieu et al., 2015), to product licensure, as outlined previously (Kaslow, 2002; WHO Publication, 2000). With recognition of the role of TBV in malaria elimination and growing funding support toward TBV development, we anticipate significant progress will be made in the coming years.

ACKNOWLEDGEMENTS

The authors thank Drs Nicholas MacDonald, David Narum for sharing unpublished data. YW was supported by Division of Intramural Research, National Institute of Allergy and Infectious Diseases. TSC was supported by the PATH Malaria Vaccine Initiative. TT is supported in part by MEXT KAKENHI (23117008), and JSPS KAKENHI (26253026 and 26670202), Japan.

REFERENCES

Agnandji, S.T., Lell, B., Soulanoudjingar, S.S., Fernandes, J.F., Abossolo, B.P., Conzelmann, C., RTS, S.C.T.P., 2011. First results of phase 3 trial of RTS,S/AS01 malaria vaccine in African children. N. Engl. J. Med. 365 (20), 1863−1875. http://dx.doi.org/10.1056/NEJMoa1102287.

Alavi, Y., Arai, M., Mendoza, J., Tufet-Bayona, M., Sinha, R., Fowler, K., Sinden, R.E., 2003. The dynamics of interactions between *Plasmodium* and the mosquito: a study of the infectivity of *Plasmodium berghei* and *Plasmodium gallinaceum*, and their transmission by *Anopheles stephensi*, *Anopheles gambiae* and *Aedes aegypti*. Int. J. Parasitol. 33 (9), 933−943.

Almeida, A.P., Billingsley, P.F., 2002. Induced immunity against the mosquito *Anopheles stephensi* (Diptera: Culicidae): effects of cell fraction antigens on survival, fecundity, and *Plasmodium berghei* (Eucoccidiida: Plasmodiidae) transmission. J. Med. Entomol. 39 (1), 207−214.

Alonso, P.L., Brown, G., Arevalo-Herrera, M., Binka, F., Chitnis, C., Collins, F., Tanner, M., 2011. A research agenda to underpin malaria eradication. PLoS Med. 8 (1), e1000406. http://dx.doi.org/10.1371/journal.pmed.1000406.

Arevalo-Herrera, M., Solarte, Y., Rocha, L., Alvarez, D., Beier, J.C., Herrera, S., 2011. Characterization of *Plasmodium vivax* transmission-blocking activity in low to moderate malaria transmission settings of the Colombian Pacific coast. Am. J. Trop. Med. Hyg. 84 (2 Suppl.), 71−77. http://dx.doi.org/10.4269/ajtmh.2011.10-0085.

Armistead, J.S., Morlais, I., Mathias, D.K., Jardim, J.G., Joy, J., Fridman, A., Dinglasan, R.R., 2014. Antibodies to a single, conserved epitope in Anopheles APN1 inhibit universal transmission of *Plasmodium falciparum* and *Plasmodium vivax* malaria. Infect. Immun. 82 (2), 818−829. http://dx.doi.org/10.1128/IAI.01222-13.

Aucouturier, J., Dupuis, L., Deville, S., Ascarateil, S., Ganne, V., 2002. Montanide ISA 720 and 51: a new generation of water in oil emulsions as adjuvants for human vaccines. Expert Rev. Vaccines 1 (1), 111−118. http://dx.doi.org/10.1586/14760584.1.1.111.

Barr, P.J., Green, K.M., Gibson, H.L., Bathurst, I.C., Quakyi, I.A., Kaslow, D.C., 1991. Recombinant Pfs25 protein of *Plasmodium falciparum* elicits malaria transmission-blocking immunity in experimental animals. J. Exp. Med. 174 (5), 1203–1208.

Beier, J.C., Copeland, R.S., Mtalib, R., Vaughan, J.A., 1992. Ookinete rates in Afrotropical anopheline mosquitoes as a measure of human malaria infectiousness. Am. J. Trop. Med. Hyg. 47 (1), 41–46.

Bergmann-Leitner, E.S., Mease, R.M., De La Vega, P., Savranskaya, T., Polhemus, M., Ockenhouse, C., Angov, E., 2010. Immunization with pre-erythrocytic antigen CelTOS from *Plasmodium falciparum* elicits cross-species protection against heterologous challenge with *Plasmodium berghei*. PLoS One 5 (8), e12294. http://dx.doi.org/10.1371/journal.pone.0012294.

Billingsley, P.F., Medley, G.F., Charlwood, D., Sinden, R.E., 1994. Relationship between prevalence and intensity of *Plasmodium falciparum* infection in natural populations of Anopheles mosquitoes. Am. J. Trop. Med. Hyg. 51 (3), 260–270.

Blagborough, A.M., Churcher, T.S., Upton, L.M., Ghani, A.C., Gething, P.W., Sinden, R.E., 2013a. Transmission-blocking interventions eliminate malaria from laboratory populations. Nat. Commun. 4, 1812. http://dx.doi.org/10.1038/ncomms2840.

Blagborough, A.M., Delves, M.J., Ramakrishnan, C., Lal, K., Butcher, G., Sinden, R.E., 2013b. Assessing transmission blockade in *Plasmodium* spp. Methods Mol. Biol. 923, 577–600. http://dx.doi.org/10.1007/978-1-62703-026-7_40.

Blagborough, A.M., Sinden, R.E., 2009. *Plasmodium berghei* HAP2 induces strong malaria transmission-blocking immunity in vivo and in vitro. Vaccine 27 (38), 5187–5194. http://dx.doi.org/10.1016/j.vaccine.2009.06.069.

Blagborough, A.M., Yoshida, S., Sattabongkot, J., Tsuboi, T., Sinden, R.E., 2010. Intranasal and intramuscular immunization with Baculovirus Dual Expression System-based Pvs25 vaccine substantially blocks *Plasmodium vivax* transmission. Vaccine 28 (37), 6014–6020. http://dx.doi.org/10.1016/j.vaccine.2010.06.100.

Blanco, A.R., Paez, A., Gerold, P., Dearsly, A.L., Margos, G., Schwarz, R.T., Sinden, R.E., 1999. The biosynthesis and post-translational modification of Pbs21 an ookinete-surface protein of *Plasmodium berghei*. Mol. Biochem. Parasitol. 98 (2), 163–173.

Bonnet, S., Gouagna, C., Safeukui, I., Meunier, J.Y., Boudin, C., 2000. Comparison of artificial membrane feeding with direct skin feeding to estimate infectiousness of *Plasmodium falciparum* gametocyte carriers to mosquitoes. Trans. R. Soc. Trop. Med. Hyg. 94 (1), 103–106.

Bonnet, S., Prevot, G., Jacques, J.C., Boudin, C., Bourgouin, C., 2001. Transcripts of the malaria vector *Anopheles gambiae* that are differentially regulated in the midgut upon exposure to invasive stages of *Plasmodium falciparum*. Cell. Microbiol. 3 (7), 449–458.

Bousema, T., Dinglasan, R.R., Morlais, I., Gouagna, L.C., van Warmerdam, T., AwonoAmbene, P.H., Churcher, T.S., 2012. Mosquito feeding assays to determine the infectiousness of naturally infected *Plasmodium falciparum* gametocyte carriers. PLoS One 7 (8), e42821. http://dx.doi.org/10.1371/journal.pone.0042821.

Bousema, T., Okell, L., Shekalaghe, S., Griffin, J.T., Omar, S., Sawa, P., Drakeley, C., 2010a. Revisiting the circulation time of *Plasmodium falciparum* gametocytes: molecular detection methods to estimate the duration of gametocyte carriage and the effect of gametocytocidal drugs. Malar. J. 9, 136. http://dx.doi.org/10.1186/1475-2875-9-136.

Bousema, T., Roeffen, W., Meijerink, H., Mwerinde, H., Mwakalinga, S., van Gemert, G.J., Drakeley, C., 2010b. The dynamics of naturally acquired immune responses to *Plasmodium falciparum* sexual stage antigens Pfs230 & Pfs48/45 in a low endemic area in Tanzania. PLoS One 5 (11), e14114. http://dx.doi.org/10.1371/journal.pone.0014114.

Brooks, A., Nunes, J.K., Garnett, A., Biellik, R., Leboulleux, D., Birkett, A.J., Loucq, C., 2012. Aligning new interventions with developing country health systems: target product profiles, presentation, and clinical trial design. Glob. Public Health 7 (9), 931—945. http://dx.doi.org/10.1080/17441692.2012.699088.

Buttery, J.P., Lambert, S.B., Grimwood, K., Nissen, M.D., Field, E.J., Macartney, K.K., Kirkwood, C.D., 2011. Reduction in rotavirus-associated acute gastroenteritis following introduction of rotavirus vaccine into Australia's national childhood vaccine schedule. Pediatr. Infect. Dis. J. 30 (1 Suppl.), S25—S29. http://dx.doi.org/10.1097/INF.0b013e3181fefdee.

Carter, R., Chen, D.H., 1976. Malaria transmission blocked by immunisation with gametes of the malaria parasite. Nature 263 (5572), 57—60.

Carter, R., Coulson, A., Bhatti, S., Taylor, B.J., Elliott, J.F., 1995. Predicted disulfide-bonded structures for three uniquely related proteins of *Plasmodium falciparum*, Pfs230, Pfs48/45 and Pf12. Mol. Biochem. Parasitol. 71 (2), 203—210.

Carter, R., Graves, P.M., Keister, D.B., Quakyi, I.A., 1990. Properties of epitopes of Pfs 48/45, a target of transmission blocking monoclonal antibodies, on gametes of different isolates of *Plasmodium falciparum*. Parasite Immunol. 12 (6), 587—603.

Carter, R., Kaushal, D.C., 1984. Characterization of antigens on mosquito midgut stages of *Plasmodium gallinaceum*. III. Changes in zygote surface proteins during transformation to mature ookinete. Mol. Biochem. Parasitol. 13 (2), 235—241.

Chowdhury, D.R., Angov, E., Kariuki, T., Kumar, N., 2009. A potent malaria transmission blocking vaccine based on codon harmonized full length Pfs48/45 expressed in *Escherichia coli*. PLoS One 4 (7), e6352. http://dx.doi.org/10.1371/journal.pone.0006352.

Churcher, T.S., Blagborough, A.M., Delves, M., Ramakrishnan, C., Kapulu, M.C., Williams, A.R., Sinden, R.E., 2012. Measuring the blockade of malaria transmission—an analysis of the standard membrane feeding assay. Int. J. Parasitol. 42 (11), 1037—1044. http://dx.doi.org/10.1016/j.ijpara.2012.09.002.

Churcher, T.S., Dawes, E.J., Sinden, R.E., Christophides, G.K., Koella, J.C., Basanez, M.G., 2010. Population biology of malaria within the mosquito: density-dependent processes and potential implications for transmission-blocking interventions. Malar. J. 9, 311. http://dx.doi.org/10.1186/1475-2875-9-311.

DeFrancesco, L., 2008. Dynavax trial halted. Nat. Biotechnol. 26 (5), 484. http://dx.doi.org/10.1038/nbt0508-484a.

Delrieu, I., Leboulleux, D., Ivinson, K., Gessner, B.D., Malaria Transmission Blocking Vaccine Technical Consultation, G, 2015. Design of a phase III cluster randomized trial to assess the efficacy and safety of a malaria transmission blocking vaccine. Vaccine 33 (13), 1518—1526. http://dx.doi.org/10.1016/j.vaccine.2015.01.050.

Dessens, J.T., Beetsma, A.L., Dimopoulos, G., Wengelnik, K., Crisanti, A., Kafatos, F.C., Sinden, R.E., 1999. CTRP is essential for mosquito infection by malaria ookinetes. EMBO J. 18 (22), 6221—6227. http://dx.doi.org/10.1093/emboj/18.22.6221.

Dessens, J.T., Mendoza, J., Claudianos, C., Vinetz, J.M., Khater, E., Hassard, S., Sinden, R.E., 2001. Knockout of the rodent malaria parasite chitinase pbCHT1 reduces infectivity to mosquitoes. Infect. Immun. 69 (6), 4041—4047. http://dx.doi.org/10.1128/IAI.69.6.4041-4047.2001.

Dessens, J.T., Siden-Kiamos, I., Mendoza, J., Mahairaki, V., Khater, E., Vlachou, D., Sinden, R.E., 2003. SOAP, a novel malaria ookinete protein involved in mosquito midgut invasion and oocyst development. Mol. Microbiol. 49 (2), 319—329.

Dinglasan, R.R., Armistead, J.S., Nyland, J.F., Jiang, X., Mao, H.Q., 2013. Single-dose microparticle delivery of a malaria transmission-blocking vaccine elicits a long-lasting functional antibody response. Curr. Mol. Med. 13 (4), 479—487.

Dinglasan, R.R., Kalume, D.E., Kanzok, S.M., Ghosh, A.K., Muratova, O., Pandey, A., Jacobs-Lorena, M., 2007. Disruption of *Plasmodium falciparum* development by antibodies against a conserved mosquito midgut antigen. Proc. Natl. Acad. Sci. U.S.A. 104 (33), 13461—13466. http://dx.doi.org/10.1073/pnas.0702239104.

Doi, M., Tanabe, K., Tachibana, S., Hamai, M., Tachibana, M., Mita, T., Tsuboi, T., 2011. Worldwide sequence conservation of transmission-blocking vaccine candidate Pvs230 in *Plasmodium vivax*. Vaccine 29 (26), 4308−4315. http://dx.doi.org/10.1016/j.vaccine.2011.04.028.

Duffy, P.E., Kaslow, D.C., 1997. A novel malaria protein, Pfs28, and Pfs25 are genetically linked and synergistic as falciparum malaria transmission-blocking vaccines. Infect. Immun. 65 (3), 1109−1113.

van Dijk, M.R., Janse, C.J., Thompson, J., Waters, A.P., Braks, J.A., Dodemont, H.J., Eling, W., 2001. A central role for P48/45 in malaria parasite male gamete fertility. Cell 104 (1), 153−164.

Da, D.F., Dixit, S., Sattabonkot, J., Mu, J., Abate, L., Ramineni, B., Wu, Y., 2013. Anti-Pfs25 human plasma reduces transmission of *Plasmodium falciparum* isolates that have diverse genetic backgrounds. Infect. Immun. 81 (6), 1984−1989. http://dx.doi.org/10.1128/IAI.00016-13.

Eksi, S., Czesny, B., van Gemert, G.J., Sauerwein, R.W., Eling, W., Williamson, K.C., 2006. Malaria transmission-blocking antigen, Pfs230, mediates human red blood cell binding to exflagellating male parasites and oocyst production. Mol. Microbiol. 61 (4), 991−998. http://dx.doi.org/10.1111/j.1365-2958.2006.05284.x.

Eksi, S., Stump, A., Fanning, S.L., Shenouda, M.I., Fujioka, H., Williamson, K.C., 2002. Targeting and sequestration of truncated Pfs230 in an intraerythrocytic compartment during *Plasmodium falciparum* gametocytogenesis. Mol. Microbiol. 44 (6), 1507−1516.

Escalante, A.A., Grebert, H.M., Chaiyaroj, S.C., Riggione, F., Biswas, S., Nahlen, B.L., Lal, A.A., 2002. Polymorphism in the gene encoding the Pfs48/45 antigen of *Plasmodium falciparum*. XI. Asembo Bay Cohort Project. Mol. Biochem. Parasitol. 119 (1), 17−22.

Farrance, C.E., Chichester, J.A., Musiychuk, K., Shamloul, M., Rhee, A., Manceva, S.D., Yusibov, V., 2011a. Antibodies to plant-produced *Plasmodium falciparum* sexual stage protein Pfs25 exhibit transmission blocking activity. Hum. Vaccines (7 Suppl.), 191−198.

Farrance, C.E., Rhee, A., Jones, R.M., Musiychuk, K., Shamloul, M., Sharma, S., Yusibov, V., 2011b. A plant-produced Pfs230 vaccine candidate blocks transmission of *Plasmodium falciparum*. Clin. Vaccine Immunol. 18 (8), 1351−1357. http://dx.doi.org/10.1128/CVI.05105-11.

Fernandes, F., Teixeira, A.P., Carinhas, N., Carrondo, M.J., Alves, P.M., 2013. Insect cells as a production platform of complex virus-like particles. Expert Rev. Vaccines 12 (2), 225−236. http://dx.doi.org/10.1586/erv.12.153.

Gastanaduy, P.A., Curns, A.T., Parashar, U.D., Lopman, B.A., 2013. Gastroenteritis hospitalizations in older children and adults in the United States before and after implementation of infant rotavirus vaccination. JAMA 310 (8), 851−853. http://dx.doi.org/10.1001/jama.2013.170800.

Gerloff, D.L., Creasey, A., Maslau, S., Carter, R., 2005. Structural models for the protein family characterized by gamete surface protein Pfs230 of *Plasmodium falciparum*. Proc. Natl. Acad. Sci. U.S.A. 102 (38), 13598−13603. http://dx.doi.org/10.1073/pnas.0502378102.

Gholizadeh, S., Basseri, H.R., Zakeri, S., Ladoni, H., Djadid, N.D., 2010. Cloning, expression and transmission-blocking activity of anti-PvWARP, malaria vaccine candidate, in *Anopheles stephensi* mysorensis. Malar. J. 9, 158. http://dx.doi.org/10.1186/1475-2875-9-158.

Ghosh, A.K., Devenport, M., Jethwaney, D., Kalume, D.E., Pandey, A., Anderson, V.E., Jacobs-Lorena, M., 2009. Malaria parasite invasion of the mosquito salivary gland requires interaction between the *Plasmodium* TRAP and the *Anopheles* saglin proteins. PLoS Pathog. 5 (1), e1000265. http://dx.doi.org/10.1371/journal.ppat.1000265.

Gonzalez-Ceron, L., Alvarado-Delgado, A., Martinez-Barnetche, J., Rodriguez, M.H., Ovilla-Munoz, M., Perez, F., Villarreal-Trevino, C., 2010. Sequence variation of ookinete surface proteins Pvs25 and Pvs28 of *Plasmodium vivax* isolates from Southern Mexico and their association to local anophelines infectivity. Infect. Genet. Evol. 10 (5), 645−654. http://dx.doi.org/10.1016/j.meegid.2010.03.014.

Goodman, A.L., Blagborough, A.M., Biswas, S., Wu, Y., Hill, A.V., Sinden, R.E., Draper, S.J., 2011. A viral vectored prime-boost immunization regime targeting the malaria Pfs25 antigen induces transmission-blocking activity. PLoS One 6 (12), e29428. http://dx.doi.org/10.1371/journal.pone.0029428.

Gouagna, L.C., Yao, F., Yameogo, B., Dabire, R.K., Ouedraogo, J.B., 2013. Comparison of field-based xenodiagnosis and direct membrane feeding assays for evaluating host infectiousness to malaria vector *Anopheles gambiae*. Acta Trop. 130C, 131−139. http://dx.doi.org/10.1016/j.actatropica.2013.10.022.

Gozar, M.M., Muratova, O., Keister, D.B., Kensil, C.R., Price, V.L., Kaslow, D.C., 2001. *Plasmodium falciparum*: immunogenicity of alum-adsorbed clinical-grade TBV25-28, a yeast-secreted malaria transmission-blocking vaccine candidate. Exp. Parasitol. 97 (2), 61−69. http://dx.doi.org/10.1006/expr.2000.4580.

Graves, P.M., Burkot, T.R., Carter, R., Cattani, J.A., Lagog, M., Parker, J., Alpers, M.P., 1988a. Measurement of malarial infectivity of human populations to mosquitoes in the Madang area, Papua, New Guinea. Parasitology 96 (Pt 2), 251−263.

Graves, P.M., Carter, R., Burkot, T.R., Quakyi, I.A., Kumar, N., 1988b. Antibodies to *Plasmodium falciparum* gamete surface antigens in Papua New Guinea sera. Parasite Immunol. 10 (2), 209−218.

Gregory, A.E., Titball, R., Williamson, D., 2013. Vaccine delivery using nanoparticles. Front. Cell. Infect. Microbiol. 3, 13. http://dx.doi.org/10.3389/fcimb.2013.00013.

Grotendorst, C.A., Carter, R., 1987. Complement effects of the infectivity of *Plasmodium gallinaceum* to *Aedes aegypti* mosquitoes. II. Changes in sensitivity to complement-like factors during zygote development. J. Parasitol. 73 (5), 980−984.

Grotendorst, C.A., Kumar, N., Carter, R., Kaushal, D.C., 1984. A surface protein expressed during the transformation of zygotes of *Plasmodium gallinaceum* is a target of transmission-blocking antibodies. Infect. Immun. 45 (3), 775−777.

Gwadz, R.W., 1976. Successful immunization against the sexual stages of *Plasmodium gallinaceum*. Science 193 (4258), 1150−1151.

Hafalla, J.C., Santiago, M.L., Pasay, M.C., Ramirez, B.L., Gozar, M.M., Saul, A., Kaslow, D.C., 1997. Minimal variation in the Pfs28 ookinete antigen from Philippine field isolates of *Plasmodium falciparum*. Mol. Biochem. Parasitol. 87 (1), 97−99.

Healer, J., McGuinness, D., Hopcroft, P., Haley, S., Carter, R., Riley, E., 1997. Complement-mediated lysis of *Plasmodium falciparum* gametes by malaria-immune human sera is associated with antibodies to the gamete surface antigen Pfs230. Infect. Immun. 65 (8), 3017−3023.

Herrera, S., Fernandez, O.L., Vera, O., Cardenas, W., Ramirez, O., Palacios, R., Arevalo-Herrera, M., 2011. Phase I safety and immunogenicity trial of *Plasmodium vivax* CS derived long synthetic peptides adjuvanted with montanide ISA 720 or montanide ISA 51. Am. J. Trop. Med. Hyg. 84 (2 Suppl.), 12−20. http://dx.doi.org/10.4269/ajtmh.2011.09-0516.

Hill, A.V., Reyes-Sandoval, A., O'Hara, G., Ewer, K., Lawrie, A., Goodman, A., Draper, S.J., 2010. Prime-boost vectored malaria vaccines: progress and prospects. Hum. Vaccines 6 (1), 78−83.

Hisaeda, H., Stowers, A.W., Tsuboi, T., Collins, W.E., Sattabongkot, J.S., Suwanabun, N., Kaslow, D.C., 2000. Antibodies to malaria vaccine candidates Pvs25 and Pvs28 completely block the ability of *Plasmodium vivax* to infect mosquitoes. Infect. Immun. 68 (12), 6618−6623.

Hu, J., Chen, Z., Gu, J., Wan, M., Shen, Q., Kieny, M.P., Pan, W., 2008. Safety and immunogenicity of a malaria vaccine, *Plasmodium falciparum* AMA-1/MSP-1 chimeric protein formulated in montanide ISA 720 in healthy adults. PLoS One 3 (4), e1952. http://dx.doi.org/10.1371/journal.pone.0001952.

Huber, M., Cabib, E., Miller, L.H., 1991. Malaria parasite chitinase and penetration of the mosquito peritrophic membrane. Proc. Natl. Acad. Sci. U.S.A. 88 (7), 2807–2810.

Huff, C.G., Marchbank, D.F., Shiroishi, T., 1958. Changes in infectiousness of malarial gametocytes. II. Analysis of the possible causative factors. Exp. Parasitol. 7 (4), 399–417.

Jones, R.M., Chichester, J.A., Mett, V., Jaje, J., Tottey, S., Manceva, S., Yusibov, V., 2013. A plant-produced Pfs25 VLP malaria vaccine candidate induces persistent transmission blocking antibodies against *Plasmodium falciparum* in immunized mice. PLoS One 8 (11), e79538. http://dx.doi.org/10.1371/journal.pone.0079538.

Kadota, K., Ishino, T., Matsuyama, T., Chinzei, Y., Yuda, M., 2004. Essential role of membrane-attack protein in malarial transmission to mosquito host. Proc. Natl. Acad. Sci. U.S.A. 101 (46), 16310–16315. http://dx.doi.org/10.1073/pnas.0406187101.

Kaiser, K., Camargo, N., Coppens, I., Morrisey, J.M., Vaidya, A.B., Kappe, S.H., 2004. A member of a conserved *Plasmodium* protein family with membrane-attack complex/perforin (MACPF)-like domains localizes to the micronemes of sporozoites. Mol. Biochem. Parasitol. 133 (1), 15–26.

Kariu, T., Ishino, T., Yano, K., Chinzei, Y., Yuda, M., 2006. CelTOS, a novel malarial protein that mediates transmission to mosquito and vertebrate hosts. Mol. Microbiol. 59 (5), 1369–1379. http://dx.doi.org/10.1111/j.1365-2958.2005.05024.x.

Kaslow, D.C., 2002. Transmission-blocking vaccines. Chem. Immunol. 80, 287–307.

Kaslow, D.C., Bathurst, I.C., Isaacs, S.N., Keister, D.B., Moss, B., Barr, P.J., 1992. Induction of *Plasmodium falciparum* transmission-blocking antibodies by recombinant Pfs25. Mem. Inst. Oswaldo Cruz 87 (Suppl. 3), 175–177.

Kaslow, D.C., Bathurst, I.C., Lensen, T., Ponnudurai, T., Barr, P.J., Keister, D.B., 1994. *Saccharomyces cerevisiae* recombinant Pfs25 adsorbed to alum elicits antibodies that block transmission of *Plasmodium falciparum*. Infect. Immun. 62 (12), 5576–5580.

Kaslow, D.C., Quakyi, I.A., Keister, D.B., 1989. Minimal variation in a vaccine candidate from the sexual stage of *Plasmodium falciparum*. Mol. Biochem. Parasitol. 32 (1), 101–103.

Kaslow, D.C., Quakyi, I.A., Syin, C., Raum, M.G., Keister, D.B., Coligan, J.E., Miller, L.H., 1988. A vaccine candidate from the sexual stage of human malaria that contains EGF-like domains. Nature 333 (6168), 74–76. http://dx.doi.org/10.1038/333074a0.

Kaslow, D.C., Shiloach, J., 1994. Production, purification and immunogenicity of a malaria transmission-blocking vaccine candidate: TBV25H expressed in yeast and purified using nickel-NTA agarose. Biotechnol. N. Y. 12 (5), 494–499.

Kaushal, D.C., Carter, R., Rener, J., Grotendorst, C.A., Miller, L.H., Howard, R.J., 1983. Monoclonal antibodies against surface determinants on gametes of *Plasmodium gallinaceum* block transmission of malaria parasites to mosquitoes. J. Immunol. 131 (5), 2557–2562.

Kelly, D.F., Moxon, E.R., Pollard, A.J., 2004. *Haemophilus influenzae* type b conjugate vaccines. Immunology 113 (2), 163–174. http://dx.doi.org/10.1111/j.1365-2567.2004.01971.x.

Kim, T.S., Kim, H.H., Kim, J.Y., Kong, Y., Na, B.K., Lin, K., Lee, H.W., 2011a. Comparison of the antibody responses to *Plasmodium vivax* and *Plasmodium falciparum* antigens in residents of Mandalay, Myanmar. Malar. J. 10, 228. http://dx.doi.org/10.1186/1475-2875-10-228.

Kim, T.S., Kim, H.H., Moon, S.U., Lee, S.S., Shin, E.H., Oh, C.M., Lee, H.W., 2011b. The role of Pvs28 in sporozoite development in *Anopheles sinensis* and its longevity in BALB/c mice. Exp. Parasitol. 127 (2), 346–350. http://dx.doi.org/10.1016/j.exppara.2010.08.015.

Kocken, C.H., Jansen, J., Kaan, A.M., Beckers, P.J., Ponnudurai, T., Kaslow, D.C., Schoenmakers, J.G., 1993. Cloning and expression of the gene coding for the transmission blocking target antigen Pfs48/45 of *Plasmodium falciparum*. Mol. Biochem. Parasitol. 61 (1), 59—68.

Kubler-Kielb, J., Majadly, F., Biesova, Z., Mocca, C.P., Guo, C., Nussenzweig, R., Schneerson, R., 2010. A bicomponent *Plasmodium falciparum* investigational vaccine composed of protein-peptide conjugates. Proc. Natl. Acad. Sci. U.S.A. 107 (3), 1172—1177. http://dx.doi.org/10.1073/pnas.0913374107.

Kubler-Kielb, J., Majadly, F., Wu, Y., Narum, D.L., Guo, C., Miller, L.H., Schneerson, R., 2007. Long-lasting and transmission-blocking activity of antibodies to *Plasmodium falciparum* elicited in mice by protein conjugates of Pfs25. Proc. Natl. Acad. Sci. U.S.A. 104 (1), 293—298. http://dx.doi.org/10.1073/pnas.0609885104.

Kumar, N., 1985. Phase separation in Triton X-114 of antigens of transmission blocking immunity in *Plasmodium gallinaceum*. Mol. Biochem. Parasitol. 17 (3), 343—358.

Kumar, N., Carter, R., 1984. Biosynthesis of the target antigens of antibodies blocking transmission of *Plasmodium falciparum*. Mol. Biochem. Parasitol. 13 (3), 333—342.

Kumar, N., Carter, R., 1985. Biosynthesis of two stage-specific membrane proteins during transformation of *Plasmodium gallinaceum* zygotes into ookinetes. Mol. Biochem. Parasitol. 14 (2), 127—139.

Kumar, R., Angov, E., Kumar, N., 2014. Potent malaria transmission-blocking antibody responses elicited by *Plasmodium falciparum* Pfs25 expressed in *Escherichia coli* after Successful protein refolding. Infect. Immun. 82 (4), 1453—1459. http://dx.doi.org/10.1128/IAI.01438-13.

van der Kolk, M., De Vlas, S.J., Saul, A., van de Vegte-Bolmer, M., Eling, W.M., Sauerwein, R.W., 2005. Evaluation of the standard membrane feeding assay (SMFA) for the determination of malaria transmission-reducing activity using empirical data. Parasitology 130 (Pt 1), 13—22.

Lal, A.A., Patterson, P.S., Sacci, J.B., Vaughan, J.A., Paul, C., Collins, W.E., Azad, A.F., 2001. Anti-mosquito midgut antibodies block development of *Plasmodium falciparum* and *Plasmodium vivax* in multiple species of Anopheles mosquitoes and reduce vector fecundity and survivorship. Proc. Natl. Acad. Sci. U.S.A. 98 (9), 5228—5233. http://dx.doi.org/10.1073/pnas.091447398.

Lavazec, C., Bonnet, S., Thiery, I., Boisson, B., Bourgouin, C., 2005. cpbAg1 encodes an active carboxypeptidase B expressed in the midgut of *Anopheles gambiae*. Insect Mol. Biol. 14 (2), 163—174. http://dx.doi.org/10.1111/j.1365-2583.2004.00541.x.

Lavazec, C., Boudin, C., Lacroix, R., Bonnet, S., Diop, A., Thiberge, S., Bourgouin, C., 2007. Carboxypeptidases B of *Anopheles gambiae* as targets for a *Plasmodium falciparum* transmission-blocking vaccine. Infect. Immun. 75 (4), 1635—1642. http://dx.doi.org/10.1128/IAI.00864-06.

Lawrence, G.W., Saul, A., Giddy, A.J., Kemp, R., Pye, D., 1997. Phase I trial in humans of an oil-based adjuvant SEPPIC MONTANIDE ISA 720. Vaccine 15 (2), 176—178.

Lensen, A.H., Van Gemert, G.J., Bolmer, M.G., Meis, J.F., Kaslow, D., Meuwissen, J.H., Ponnudurai, T., 1992. Transmission blocking antibody of the *Plasmodium falciparum* zygote/ookinete surface protein Pfs25 also influences sporozoite development. Parasite Immunol. 14 (5), 471—479.

Li, F., Patra, K.P., Vinetz, J.M., 2005. An anti-chitinase malaria transmission-blocking single-chain antibody as an effector molecule for creating a *Plasmodium falciparum*-refractory mosquito. J. Infect. Dis. 192 (5), 878—887. http://dx.doi.org/10.1086/432552.

Li, F., Templeton, T.J., Popov, V., Comer, J.E., Tsuboi, T., Torii, M., Vinetz, J.M., 2004. *Plasmodium* ookinete-secreted proteins secreted through a common micronemal pathway are targets of blocking malaria transmission. J. Biol. Chem. 279 (25), 26635—26644. http://dx.doi.org/10.1074/jbc.M401385200.

Lin, F.Y., Ho, V.A., Khiem, H.B., Trach, D.D., Bay, P.V., Thanh, T.C., Szu, S.C., 2001. The efficacy of a *Salmonella typhi* Vi conjugate vaccine in two-to-five-year-old children. N. Engl. J. Med. 344 (17), 1263−1269. http://dx.doi.org/10.1056/NEJM200104263441701.

Liu, Y., Tewari, R., Ning, J., Blagborough, A.M., Garbom, S., Pei, J., Billker, O., 2008. The conserved plant sterility gene HAP2 functions after attachment of fusogenic membranes in *Chlamydomonas* and *Plasmodium* gametes. Genes Dev. 22 (8), 1051−1068. http://dx.doi.org/10.1101/gad.1656508.

Malkin, E.M., Durbin, A.P., Diemert, D.J., Sattabongkot, J., Wu, Y., Miura, K., Saul, A., 2005. Phase 1 vaccine trial of Pvs25H: a transmission blocking vaccine for *Plasmodium vivax* malaria. Vaccine 23 (24), 3131−3138. http://dx.doi.org/10.1016/j.vaccine.2004.12.019.

Margos, G., Navarette, S., Butcher, G., Davies, A., Willers, C., Sinden, R.E., Lachmann, P.J., 2001. Interaction between host complement and mosquito-midgut-stage *Plasmodium berghei*. Infect. Immun. 69 (8), 5064−5071. http://dx.doi.org/10.1128/IAI.69.8.5064-5071.2001.

Mathias, D.K., Plieskatt, J.L., Armistead, J.S., Bethony, J.M., Abdul-Majid, K.B., McMillan, A., Dinglasan, R.R., 2012. Expression, immunogenicity, histopathology, and potency of a mosquito-based malaria transmission-blocking recombinant vaccine. Infect. Immun. 80 (4), 1606−1614. http://dx.doi.org/10.1128/IAI.06212-11.

McCarthy, J.S., Marjason, J., Elliott, S., Fahey, P., Bang, G., Malkin, E., Anders, R.F., 2011. A phase 1 trial of MSP2-C1, a blood-stage malaria vaccine containing 2 isoforms of MSP2 formulated with Montanide(R) ISA 720. PLoS One 6 (9), e24413. http://dx.doi.org/10.1371/journal.pone.0024413.

Medley, G.F., Sinden, R.E., Fleck, S., Billingsley, P.F., Tirawanchai, N., Rodriguez, M.H., 1993. Heterogeneity in patterns of malarial oocyst infections in the mosquito vector. Parasitology 106 (Pt 5), 441−449.

Mendis, K.N., Munesinghe, Y.D., de Silva, Y.N., Keragalla, I., Carter, R., 1987. Malaria transmission-blocking immunity induced by natural infections of *Plasmodium vivax* in humans. Infect. Immun. 55 (2), 369−372.

Milek, R.L., DeVries, A.A., Roeffen, W.F., Stunnenberg, H., Rottier, P.J., Konings, R.N., 1998a. *Plasmodium falciparum*: heterologous synthesis of the transmission-blocking vaccine candidate Pfs48/45 in recombinant vaccinia virus-infected cells. Exp. Parasitol. 90 (2), 165−174. http://dx.doi.org/10.1006/expr.1998.4315.

Milek, R.L., Roeffen, W.F., Kocken, C.H., Jansen, J., Kaan, A.M., Eling, W.M., Konings, R.N., 1998b. Immunological properties of recombinant proteins of the transmission blocking vaccine candidate, Pfs48/45, of the human malaria parasite *Plasmodium falciparum* produced in *Escherichia coli*. Parasite Immunol. 20 (8), 377−385.

Milek, R.L., Stunnenberg, H.G., Konings, R.N., 2000. Assembly and expression of a synthetic gene encoding the antigen Pfs48/45 of the human malaria parasite *Plasmodium falciparum* in yeast. Vaccine 18 (14), 1402−1411.

Miles, A.P., Zhang, Y., Saul, A., Stowers, A.W., 2002. Large-scale purification and characterization of malaria vaccine candidate antigen Pvs25H for use in clinical trials. Protein Expr. Purif. 25 (1), 87−96. http://dx.doi.org/10.1006/prep.2001.1613.

Miura, K., Deng, B., Tullo, G., Diouf, A., Moretz, S.E., Locke, E., Long, C.A., 2013a. Qualification of standard membrane-feeding assay with *Plasmodium falciparum* malaria and potential improvements for future assays. PLoS One 8 (3), e57909. http://dx.doi.org/10.1371/journal.pone.0057909.

Miura, K., Keister, D.B., Muratova, O.V., Sattabongkot, J., Long, C.A., Saul, A., 2007. Transmission-blocking activity induced by malaria vaccine candidates Pfs25/Pvs25 is a direct and predictable function of antibody titer. Malar. J. 6, 107. http://dx.doi.org/10.1186/1475-2875-6-107.

Miura, K., Takashima, E., Deng, B., Tullo, G., Diouf, A., Moretz, S.E., Tsuboi, T., 2013b. Functional comparison of *Plasmodium falciparum* transmission-blocking vaccine candidates by the standard membrane-feeding assay. Infect. Immun. 81 (12), 4377—4382. http://dx.doi.org/10.1128/IAI.01056-13.

Miyata, T., Harakuni, T., Tsuboi, T., Sattabongkot, J., Ikehara, A., Tachibana, M., Arakawa, T., 2011. Tricomponent immunopotentiating system as a novel molecular design strategy for malaria vaccine development. Infect. Immun. 79 (10), 4260—4275. http://dx.doi.org/10.1128/IAI.05214-11.

Miyata, T., Harakuni, T., Tsuboi, T., Sattabongkot, J., Kohama, H., Tachibana, M., Arakawa, T., 2010. Plasmodium vivax ookinete surface protein Pvs25 linked to cholera toxin B subunit induces potent transmission-blocking immunity by intranasal as well as subcutaneous immunization. Infect. Immun. 78 (9), 3773—3782. http://dx.doi.org/10.1128/IAI.00306-10.

Mizutani, M., Iyori, M., Blagborough, A.M., Fukumoto, S., Funatsu, T., Sinden, R.E., Yoshida, S., 2014. Baculovirus-vectored multistage *Plasmodium vivax* vaccine induces both protective and transmission-blocking immunities against transgenic rodent malaria parasites. Infect. Immun. 82 (10), 4348—4357. http://dx.doi.org/10.1128/IAI.02040-14.

Mlambo, G., Kumar, N., Yoshida, S., 2010. Functional immunogenicity of baculovirus expressing Pfs25, a human malaria transmission-blocking vaccine candidate antigen. Vaccine 28 (43), 7025—7029. http://dx.doi.org/10.1016/j.vaccine.2010.08.022.

Molina-Cruz, A., Barillas-Mury, C., 2014. The remarkable journey of adaptation of the *Plasmodium falciparum* malaria parasite to New World anopheline mosquitoes. Mem. Inst. Oswaldo Cruz 0, 17—20.

Moorthy, V.S., Newman, R.D., Okwo-Bele, J.M., 2013. Malaria vaccine technology roadmap. Lancet 382 (9906), 1700—1701. http://dx.doi.org/10.1016/S0140-6736(13)62238-2.

Mori, T., Hirai, M., Kuroiwa, T., Miyagishima, S.Y., 2010. The functional domain of GCS1-based gamete fusion resides in the amino terminus in plant and parasite species. PLoS One 5 (12), e15957. http://dx.doi.org/10.1371/journal.pone.0015957.

Niederwieser, I., Felger, I., Beck, H.P., 2001. Limited polymorphism in *Plasmodium falciparum* sexual-stage antigens. Am. J. Trop. Med. Hyg. 64 (1—2), 9—11.

Nolz, J.C., Harty, J.T., 2011. Strategies and implications for prime-boost vaccination to generate memory CD8 T cells. Adv. Exp. Med. Biol. 780, 69—83. http://dx.doi.org/10.1007/978-1-4419-5632-3_7.

Okulate, M.A., Kalume, D.E., Reddy, R., Kristiansen, T., Bhattacharyya, M., Chaerkady, R., Kumar, N., 2007. Identification and molecular characterization of a novel protein Saglin as a target of monoclonal antibodies affecting salivary gland infectivity of *Plasmodium* sporozoites. Insect Mol. Biol. 16 (6), 711—722. http://dx.doi.org/10.1111/j.1365-2583.2007.00765.x.

Oliveira, G.A., Wetzel, K., Calvo-Calle, J.M., Nussenzweig, R., Schmidt, A., Birkett, A., Nardin, E.H., 2005. Safety and enhanced immunogenicity of a hepatitis B core particle *Plasmodium falciparum* malaria vaccine formulated in adjuvant Montanide ISA 720 in a phase I trial. Infect. Immun. 73 (6), 3587—3597. http://dx.doi.org/10.1128/IAI.73.6.3587-3597.2005.

Olotu, A., Fegan, G., Wambua, J., Nyangweso, G., Awuondo, K.O., Leach, A., Bejon, P., 2013. Four-year efficacy of RTS,S/AS01E and its interaction with malaria exposure. N. Engl. J. Med. 368 (12), 1111—1120. http://dx.doi.org/10.1056/NEJMoa1207564.

Ong, C.S., Zhang, K.Y., Eida, S.J., Graves, P.M., Dow, C., Looker, M., Targett, G.A., 1990. The primary antibody response of malaria patients to *Plasmodium falciparum* sexual stage antigens which are potential transmission blocking vaccine candidates. Parasite Immunol. 12 (5), 447—456.

Ouedraogo, A.L., Roeffen, W., Luty, A.J., de Vlas, S.J., Nebie, I., Ilboudo-Sanogo, E., Sauerwein, R., 2011. Naturally acquired immune responses to *Plasmodium falciparum*

sexual stage antigens Pfs48/45 and Pfs230 in an area of seasonal transmission. Infect. Immun. 79 (12), 4957—4964. http://dx.doi.org/10.1128/IAI.05288-11.
Outchkourov, N.S., Roeffen, W., Kaan, A., Jansen, J., Luty, A., Schuiffel, D., Stunnenberg, H.G., 2008. Correctly folded Pfs48/45 protein of *Plasmodium falciparum* elicits malaria transmission-blocking immunity in mice. Proc. Natl. Acad. Sci. U.S.A. 105 (11), 4301—4305. http://dx.doi.org/10.1073/pnas.0800459105.
Passwell, J.H., Ashkenazi, S., Harlev, E., Miron, D., Ramon, R., Farzam, N., Israel Shigella Study, G, 2003. Safety and immunogenicity of *Shigella sonnei*-CRM9 and *Shigella flexneri* type 2a-rEPAsucc conjugate vaccines in one- to four-year-old children. Pediatr. Infect. Dis. J. 22 (8), 701—706. http://dx.doi.org/10.1097/01.inf.0000078156.03697.a5.
Passwell, J.H., Ashkenzi, S., Banet-Levi, Y., Ramon-Saraf, R., Farzam, N., Lerner-Geva, L., Israeli Shigella Study, G, 2010. Age-related efficacy of Shigella O-specific polysaccharide conjugates in 1-4-year-old Israeli children. Vaccine 28 (10), 2231—2235. http://dx.doi.org/10.1016/j.vaccine.2009.12.050.
Pichichero, M.E., 2013. Protein carriers of conjugate vaccines: characteristics, development and clinical trials. Hum. Vaccines Immunother. 9 (12).
Ponsa, N., Sattabongkot, J., Kittayapong, P., Eikarat, N., Coleman, R.E., 2003. Transmission-blocking activity of tafenoquine (WR-238605) and artelinic acid against naturally circulating strains of *Plasmodium vivax* in Thailand. Am. J. Trop. Med. Hyg. 69 (5), 542—547.
Premawansa, S., Gamage-Mendis, A., Perera, L., Begarnie, S., Mendis, K., Carter, R., 1994. *Plasmodium falciparum* malaria transmission-blocking immunity under conditions of low endemicity as in Sri Lanka. Parasite Immunol. 16 (1), 35—42.
Qian, F., Aebig, J.A., Reiter, K., Barnafo, E., Zhang, Y., Shimp Jr., R.L., Wu, Y., 2009. Enhanced antibody responses to *Plasmodium falciparum* Pfs28 induced in mice by conjugation to ExoProtein A of *Pseudomonas aeruginosa* with an improved procedure. Microbes Infect. 11 (3), 408—412. http://dx.doi.org/10.1016/j.micinf.2008.12.009.
Qian, F., Rausch, K.M., Muratova, O., Zhou, H., Song, G., Diouf, A., Mullen, G.E., 2008. Addition of CpG ODN to recombinant *Pseudomonas aeruginosa* ExoProtein A conjugates of AMA1 and Pfs25 greatly increases the number of responders. Vaccine 26 (20), 2521—2527. http://dx.doi.org/10.1016/j.vaccine.2008.03.005.
Qian, F., Wu, Y., Muratova, O., Zhou, H., Dobrescu, G., Duggan, P., Mullen, G.E., 2007. Conjugating recombinant proteins to *Pseudomonas aeruginosa* ExoProtein A: a strategy for enhancing immunogenicity of malaria vaccine candidates. Vaccine 25 (20), 3923—3933. http://dx.doi.org/10.1016/j.vaccine.2007.02.073.
Quakyi, I.A., Carter, R., Rener, J., Kumar, N., Good, M.F., Miller, L.H., 1987. The 230-kDa gamete surface protein of *Plasmodium falciparum* is also a target for transmission-blocking antibodies. J. Immunol. 139 (12), 4213—4217.
Ramakrishnan, C., Dessens, J.T., Armson, R., Pinto, S.B., Talman, A.M., Blagborough, A.M., Sinden, R.E., 2011. Vital functions of the malarial ookinete protein, CTRP, reside in the A domains. Int. J. Parasitol. 41 (10), 1029—1039. http://dx.doi.org/10.1016/j.ijpara.2011.05.007.
Ramasamy, M.S., Ramasamy, R., 1990. Effect of anti-mosquito antibodies on the infectivity of the rodent malaria parasite *Plasmodium berghei* to *Anopheles farauti*. Med. Vet. Entomol. 4 (2), 161—166.
Ramjanee, S., Robertson, J.S., Franke-Fayard, B., Sinha, R., Waters, A.P., Janse, C.J., Sinden, R.E., 2007. The use of transgenic *Plasmodium berghei* expressing the *Plasmodium vivax* antigen P25 to determine the transmission-blocking activity of sera from malaria vaccine trials. Vaccine 25 (5), 886—894. http://dx.doi.org/10.1016/j.vaccine.2006.09.035.
Rashid, H., Khandaker, G., Booy, R., 2012. Vaccination and herd immunity: what more do we know? Curr. Opin. Infect. Dis. 25 (3), 243—249. http://dx.doi.org/10.1097/QCO.0b013e328352f727.

Read, D., Lensen, A.H., Begarnie, S., Haley, S., Raza, A., Carter, R., 1994. Transmission-blocking antibodies against multiple, non-variant target epitopes of the *Plasmodium falciparum* gamete surface antigen Pfs230 are all complement-fixing. Parasite Immunol. 16 (10), 511–519.

Rener, J., Graves, P.M., Carter, R., Williams, J.L., Burkot, T.R., 1983. Target antigens of transmission-blocking immunity on gametes of *Plasmodium falciparum*. J. Exp. Med. 158 (3), 976–981.

Richards, J.S., MacDonald, N.J., Eisen, D.P., 2006. Limited polymorphism in *Plasmodium falciparum* ookinete surface antigen, von Willebrand factor A domain-related protein from clinical isolates. Malar. J. 5, 55. http://dx.doi.org/10.1186/1475-2875-5-55.

Riley, E.M., Bennett, S., Jepson, A., Hassan-King, M., Whittle, H., Olerup, O., Carter, R., 1994. Human antibody responses to Pfs 230, a sexual stage-specific surface antigen of *Plasmodium falciparum*: non-responsiveness is a stable phenotype but does not appear to be genetically regulated. Parasite Immunol. 16 (2), 55–62.

Roberts, L., Enserink, M., 2007. Malaria. Did they really say... eradication? Science 318 (5856), 1544–1545. http://dx.doi.org/10.1126/science.318.5856.1544.

Roeffen, W., Geeraedts, F., Eling, W., Beckers, P., Wizel, B., Kumar, N., Sauerwein, R., 1995. Transmission blockade of *Plasmodium falciparum* malaria by anti-Pfs230-specific antibodies is isotype dependent. Infect. Immun. 63 (2), 467–471.

Roeffen, W., Teelen, K., van As, J., vd Vegte-Bolmer, M., Eling, W., Sauerwein, R., 2001. *Plasmodium falciparum*: production and characterization of rat monoclonal antibodies specific for the sexual-stage Pfs48/45 antigen. Exp. Parasitol. 97 (1), 45–49. http://dx.doi.org/10.1006/expr.2000.4586.

Rypniewski, W.R., Perrakis, A., Vorgias, C.E., Wilson, K.S., 1994. Evolutionary divergence and conservation of trypsin. Protein Eng. 7 (1), 57–64.

Sattabongkot, J., Tsuboi, T., Hisaeda, H., Tachibana, M., Suwanabun, N., Rungruang, T., Torii, M., 2003. Blocking of transmission to mosquitoes by antibody to *Plasmodium vivax* malaria vaccine candidates Pvs25 and Pvs28 despite antigenic polymorphism in field isolates. Am. J. Trop. Med. Hyg. 69 (5), 536–541.

Saul, A., 2008. Efficacy model for mosquito stage transmission blocking vaccines for malaria. Parasitology 135 (13), 1497–1506. http://dx.doi.org/10.1017/S0031182008000280.

Saul, A., Fay, M.P., 2007. Human immunity and the design of multi-component, single target vaccines. PLoS One 2 (9), e850. http://dx.doi.org/10.1371/journal.pone.0000850.

Saul, A., Hensmann, M., Sattabongkot, J., Collins, W.E., Barnwell, J.W., Langermans, J.A., Thomas, A.W., 2007. Immunogenicity in rhesus of the *Plasmodium vivax* mosquito stage antigen Pvs25H with Alhydrogel and Montanide ISA 720. Parasite Immunol. 29 (10), 525–533. http://dx.doi.org/10.1111/j.1365-3024.2007.00971.x.

Saul, A., Lawrence, G., Allworth, A., Elliott, S., Anderson, K., Rzepczyk, C., Anders, R.F., 2005. A human phase 1 vaccine clinical trial of the *Plasmodium falciparum* malaria vaccine candidate apical membrane antigen 1 in Montanide ISA720 adjuvant. Vaccine 23 (23), 3076–3083. http://dx.doi.org/10.1016/j.vaccine.2004.09.040.

Saul, A., Lawrence, G., Smillie, A., Rzepczyk, C.M., Reed, C., Taylor, D., Sturchler, D., 1999. Human phase I vaccine trials of 3 recombinant asexual stage malaria antigens with Montanide ISA720 adjuvant. Vaccine 17 (23–24), 3145–3159.

Saxena, A.K., Singh, K., Su, H.P., Klein, M.M., Stowers, A.W., Saul, A.J., Garboczi, D.N., 2006. The essential mosquito-stage P25 and P28 proteins from *Plasmodium* form tile-like triangular prisms. Nat. Struct. Mol. Biol. 13 (1), 90–91. http://dx.doi.org/10.1038/nsmb1024.

Schwartz, L., Brown, G.V., Genton, B., Moorthy, V.S., 2012. A review of malaria vaccine clinical projects based on the WHO rainbow table. Malar. J. 11, 11. http://dx.doi.org/10.1186/1475-2875-11-11.

Shahabuddin, M., Kaidoh, T., Aikawa, M., Kaslow, D.C., 1995. *Plasmodium gallinaceum*: mosquito peritrophic matrix and the parasite-vector compatibility. Exp. Parasitol. 81 (3), 386–393. http://dx.doi.org/10.1006/expr.1995.1129.

Shahabuddin, M., Lemos, F.J., Kaslow, D.C., Jacobs-Lorena, M., 1996. Antibody-mediated inhibition of *Aedes aegypti* midgut trypsins blocks sporogonic development of *Plasmodium gallinaceum*. Infect. Immun. 64 (3), 739–743.

Sharma, B., 2008. Structure and mechanism of a transmission blocking vaccine candidate protein Pfs25 from *P. falciparum*: a molecular modeling and docking study. In Silico Biol. 8 (3–4), 193–206.

Sharma, B., Jaiswal, M.K., Saxena, A.K., 2009. EGF domain II of protein Pb28 from *Plasmodium berghei* interacts with monoclonal transmission blocking antibody 13.1. J. Mol. Model 15 (4), 369–382. http://dx.doi.org/10.1007/s00894-008-0404-y.

Shi, Y.P., Alpers, M.P., Povoa, M.M., Lal, A.A., 1992. Single amino acid variation in the ookinete vaccine antigen from field isolates of *Plasmodium falciparum*. Mol. Biochem. Parasitol. 50 (1), 179–180.

Shimp Jr., R.L., Rowe, C., Reiter, K., Chen, B., Nguyen, V., Aebig, J., Narum, D.L., 2013. Development of a Pfs25-EPA malaria transmission blocking vaccine as a chemically conjugated nanoparticle. Vaccine 31 (28), 2954–2962. http://dx.doi.org/10.1016/j.vaccine.2013.04.034.

Sinden, R.E., 2010. A biologist's perspective on malaria vaccine development. Hum. Vaccines 6 (1), 3–11.

Sinden, R.E., Dawes, E.J., Alavi, Y., Waldock, J., Finney, O., Mendoza, J., Basanez, M.G., 2007. Progression of *Plasmodium berghei* through *Anopheles stephensi* is density-dependent. PLoS Pathog. 3 (12), e195. http://dx.doi.org/10.1371/journal.ppat.0030195.

Sinden, R.E., Winger, L., Carter, E.H., Hartley, R.H., Tirawanchai, N., Davies, C.S., Sluiters, J.F., 1987. Ookinete antigens of *Plasmodium berghei*: a light and electron-microscope immunogold study of expression of the 21 kDa determinant recognized by a transmission-blocking antibody. Proc. R. Soc. Lond. B Biol. Sci. 230 (1261), 443–458.

Smith, D.L., McKenzie, F.E., Snow, R.W., Hay, S.I., 2007. Revisiting the basic reproductive number for malaria and its implications for malaria control. PLoS Biol. 5 (3), e42. http://dx.doi.org/10.1371/journal.pbio.0050042.

Smith, T.A., Chitnis, N., Briet, O.J., Tanner, M., 2011. Uses of mosquito-stage transmission-blocking vaccines against *Plasmodium falciparum*. Trends Parasitol. 27 (5), 190–196. http://dx.doi.org/10.1016/j.pt.2010.12.011.

Srikrishnaraj, K.A., Ramasamy, R., Ramasamy, M.S., 1995. Antibodies to *Anopheles* midgut reduce vector competence for *Plasmodium vivax* malaria. Med. Vet. Entomol. 9 (4), 353–357.

Stone, W.J., Eldering, M., van Gemert, G.J., Lanke, K.H., Grignard, L., van de Vegte-Bolmer, M.G., Bousema, T., 2013. The relevance and applicability of oocyst prevalence as a read-out for mosquito feeding assays. Sci. Rep. 3, 3418. http://dx.doi.org/10.1038/srep03418.

Su, X., Hayton, K., Wellems, T.E., 2007. Genetic linkage and association analyses for trait mapping in *Plasmodium falciparum*. Nat. Rev. Genet. 8 (7), 497–506. http://dx.doi.org/10.1038/nrg2126.

Tachibana, M., Sato, C., Otsuki, H., Sattabongkot, J., Kaneko, O., Torii, M., Tsuboi, T., 2012. *Plasmodium vivax* gametocyte protein Pvs230 is a transmission-blocking vaccine candidate. Vaccine 30 (10), 1807–1812. http://dx.doi.org/10.1016/j.vaccine.2012.01.003.

Tachibana, M., Wu, Y., Iriko, H., Muratova, O., MacDonald, N.J., Sattabongkot, J., Tsuboi, T., 2011. N-terminal prodomain of Pfs230 synthesized using a cell-free system is sufficient to induce complement-dependent malaria transmission-blocking activity. Clin. Vaccine Immunol. 18 (8), 1343–1350. http://dx.doi.org/10.1128/CVI.05104-11.

Takeo, S., Hisamori, D., Matsuda, S., Vinetz, J., Sattabongkot, J., Tsuboi, T., 2009. Enzymatic characterization of the *Plasmodium vivax* chitinase, a potential malaria transmission-blocking target. Parasitol. Int. 58 (3), 243—248. http://dx.doi.org/10.1016/j.parint.2009.05.002.

Talwar, G.P., Singh, O., Pal, R., Chatterjee, N., Sahai, P., Dhall, K., et al., 1994. A vaccine that prevents pregnancy in women. Proc. Natl. Acad. Sci. U.S.A. 91 (18), 8532—8536.

Templeton, T.J., Kaslow, D.C., 1999. Identification of additional members define a *Plasmodium falciparum* gene superfamily which includes Pfs48/45 and Pfs230. Mol. Biochem. Parasitol. 101 (1—2), 223—227.

Theisen, M., Roeffen, W., Singh, S.K., Andersen, G., Amoah, L., van de Vegte-Bolmer, M., Sauerwein, R., 2014. A multi-stage malaria vaccine candidate targeting both transmission and asexual parasite life-cycle stages. Vaccine 32 (22), 2623—2630. http://dx.doi.org/10.1016/j.vaccine.2014.03.020.

Thiem, V.D., Lin, F.Y., Canh do, G., Son, N.H., Anh, D.D., Mao, N.D., Szu, S.C., 2011. The Vi conjugate typhoid vaccine is safe, elicits protective levels of IgG anti-Vi, and is compatible with routine infant vaccines. Clin. Vaccine Immunol. 18 (5), 730—735. http://dx.doi.org/10.1128/CVI.00532-10.

Tine, J.A., Lanar, D.E., Smith, D.M., Wellde, B.T., Schultheiss, P., Ware, L.A., Paoletti, E., 1996. NYVAC-Pf7: a poxvirus-vectored, multiantigen, multistage vaccine candidate for *Plasmodium falciparum* malaria. Infect. Immun. 64 (9), 3833—3844.

Tomas, A.M., Margos, G., Dimopoulos, G., van Lin, L.H., de Koning-Ward, T.F., Sinha, R., Sinden, R.E., 2001. P25 and P28 proteins of the malaria ookinete surface have multiple and partially redundant functions. EMBO J. 20 (15), 3975—3983. http://dx.doi.org/10.1093/emboj/20.15.3975.

Toure, Y.T., Doumbo, O., Toure, A., Bagayoko, M., Diallo, M., Dolo, A., Kaslow, D.C., 1998. Gametocyte infectivity by direct mosquito feeds in an area of seasonal malaria transmission: implications for Bancoumana, Mali as a transmission-blocking vaccine site. Am. J. Trop. Med. Hyg. 59 (3), 481—486.

Tsai, C.W., Duggan, P.F., Shimp Jr., R.L., Miller, L.H., Narum, D.L., 2006. Overproduction of *Pichia pastoris* or *Plasmodium falciparum* protein disulfide isomerase affects expression, folding and O-linked glycosylation of a malaria vaccine candidate expressed in *P. pastoris*. J. Biotechnol. 121 (4), 458—470. http://dx.doi.org/10.1016/j.jbiotec.2005.08.025.

Tsai, Y.L., Hayward, R.E., Langer, R.C., Fidock, D.A., Vinetz, J.M., 2001. Disruption of *Plasmodium falciparum* chitinase markedly impairs parasite invasion of mosquito midgut. Infect. Immun. 69 (6), 4048—4054. http://dx.doi.org/10.1128/IAI.69.6.4048-4054.2001.

Tsuboi, T., Kaneko, O., Cao, Y.M., Tachibana, M., Yoshihiro, Y., Nagao, T., Torii, M., 2004. A rapid genotyping method for the vivax malaria transmission-blocking vaccine candidates, Pvs25 and Pvs28. Parasitol. Int. 53 (3), 211—216.

Tsuboi, T., Kaslow, D.C., Gozar, M.M., Tachibana, M., Cao, Y.M., Torii, M., 1998. Sequence polymorphism in two novel *Plasmodium vivax* ookinete surface proteins, Pvs25 and Pvs28, that are malaria transmission-blocking vaccine candidates. Mol. Med. 4 (12), 772—782.

Vaughan, J.A., Noden, B.H., Beier, J.C., 1992. Population dynamics of *Plasmodium falciparum* sporogony in laboratory-infected *Anopheles gambiae*. J. Parasitol. 78 (4), 716—724.

Vermeulen, A.N., Ponnudurai, T., Beckers, P.J., Verhave, J.P., Smits, M.A., Meuwissen, J.H., 1985a. Sequential expression of antigens on sexual stages of *Plasmodium falciparum* accessible to transmission-blocking antibodies in the mosquito. J. Exp. Med. 162 (5), 1460—1476.

Vermeulen, A.N., Roeffen, W.F., Henderik, J.B., Ponnudurai, T., Beckers, P.J., Meuwissen, J.H., 1985b. *Plasmodium falciparum* transmission blocking monoclonal antibodies recognize monovalently expressed epitopes. Dev. Biol. Stand. 62, 91—97.

Vermeulen, A.N., van Deursen, J., Brakenhoff, R.H., Lensen, T.H., Ponnudurai, T., Meuwissen, J.H., 1986. Characterization of *Plasmodium falciparum* sexual stage antigens

and their biosynthesis in synchronised gametocyte cultures. Mol. Biochem. Parasitol. 20 (2), 155—163.
Vinetz, J.M., Valenzuela, J.G., Specht, C.A., Aravind, L., Langer, R.C., Ribeiro, J.M., Kaslow, D.C., 2000. Chitinases of the avian malaria parasite *Plasmodium gallinaceum*, a class of enzymes necessary for parasite invasion of the mosquito midgut. J. Biol. Chem. 275 (14), 10331—10341.
Waye, A., Jacobs, P., Tan, B., 2013. The impact of the universal infant varicella immunization strategy on Canadian varicella-related hospitalization rates. Vaccine 31 (42), 4744—4748. http://dx.doi.org/10.1016/j.vaccine.2013.08.022.
Wenger, E.A., Eckhoff, P.A., 2013. A mathematical model of the impact of present and future malaria vaccines. Malar. J. 12, 126. http://dx.doi.org/10.1186/1475-2875-12-126.
Williams, A.R., Zakutansky, S.E., Miura, K., Dicks, M.D., Churcher, T.S., Jewell, K.E., Biswas, S., 2013. Immunisation against a serine protease inhibitor reduces intensity of *Plasmodium berghei* infection in mosquitoes. Int. J. Parasitol. 43 (11), 869—874. http://dx.doi.org/10.1016/j.ijpara.2013.06.004.
Williamson, K.C., Criscio, M.D., Kaslow, D.C., 1993. Cloning and expression of the gene for *Plasmodium falciparum* transmission-blocking target antigen, Pfs230. Mol. Biochem. Parasitol. 58 (2), 355—358.
Williamson, K.C., Fujioka, H., Aikawa, M., Kaslow, D.C., 1996. Stage-specific processing of Pfs230, a *Plasmodium falciparum* transmission-blocking vaccine candidate. Mol. Biochem. Parasitol. 78 (1—2), 161—169.
Williamson, K.C., Kaslow, D.C., 1993. Strain polymorphism of *Plasmodium falciparum* transmission-blocking target antigen Pfs230. Mol. Biochem. Parasitol. 62 (1), 125—127.
Williamson, K.C., Keister, D.B., Muratova, O., Kaslow, D.C., 1995. Recombinant Pfs230, a *Plasmodium falciparum* gametocyte protein, induces antisera that reduce the infectivity of *Plasmodium falciparum* to mosquitoes. Mol. Biochem. Parasitol. 75 (1), 33—42.
Woo, M.K., Kim, K.A., Kim, J., Oh, J.S., Han, E.T., An, S.S., Lim, C.S., 2013. Sequence polymorphisms in Pvs48/45 and Pvs47 gametocyte and gamete surface proteins in *Plasmodium vivax* isolated in Korea. Mem. Inst. Oswaldo Cruz 108 (3). http://dx.doi.org/10.1590/S0074-02762013000300015.
Wu, Y., Craig, A., 2006. Comparative proteomic analysis of metabolically labelled proteins from *Plasmodium falciparum* isolates with different adhesion properties. Malar. J. 5, 67. http://dx.doi.org/10.1186/1475-2875-5-67.
Wu, Y., Ellis, R.D., Shaffer, D., Fontes, E., Malkin, E.M., Mahanty, S., Durbin, A.P., 2008. Phase 1 trial of malaria transmission blocking vaccine candidates Pfs25 and Pvs25 formulated with montanide ISA 51. PLoS One 3 (7), e2636. http://dx.doi.org/10.1371/journal.pone.0002636.
Wu, Y., Przysiecki, C., Flanagan, E., Bello-Irizarry, S.N., Ionescu, R., Muratova, O., Miller, L.H., 2006. Sustained high-titer antibody responses induced by conjugating a malarial vaccine candidate to outer-membrane protein complex. Proc. Natl. Acad. Sci. U.S.A. 103 (48), 18243—18248. http://dx.doi.org/10.1073/pnas.0608545103.
WHO Publication, 2000. Malaria Transmission Blocking Vaccines: An Ideal Public Good. http://www.who.int/tdr/publications/tdr-research-publications/malaria-transmission-blocking-vaccines/en/.
WHO Malaria Report, 2014. http://www.who.int/malaria/publications/world_malaria_report_2014/report/en/.
Yuda, M., Yano, K., Tsuboi, T., Torii, M., Chinzei, Y., 2001. von Willebrand Factor A domain-related protein, a novel microneme protein of the malaria ookinete highly conserved throughout *Plasmodium* parasites. Mol. Biochem. Parasitol. 116 (1), 65—72.
Zakeri, S., Razavi, S., Djadid, N.D., 2009. Genetic diversity of transmission blocking vaccine candidate (Pvs25 and Pvs28) antigen in *Plasmodium vivax* clinical isolates from Iran. Acta Trop. 109 (3), 176—180. http://dx.doi.org/10.1016/j.actatropica.2008.09.012.

Zieler, H., Garon, C.F., Fischer, E.R., Shahabuddin, M., 2000. A tubular network associated with the brush-border surface of the *Aedes aegypti* midgut: implications for pathogen transmission by mosquitoes. J. Exp. Biol. 203 (Pt 10), 1599–1611.

Zou, L., Miles, A.P., Wang, J., Stowers, A.W., 2003. Expression of malaria transmission-blocking vaccine antigen Pfs25 in *Pichia pastoris* for use in human clinical trials. Vaccine 21 (15), 1650–1657.

INDEX

Note: Page numbers followed by "b" and"f" indicate boxes and figures respectively.

A

A. gambiae aminopeptidase (AgAPN1), 120
Aberrant free-living existence, 52–53
  echinostomes, 53–54
  unencysted forms of *C. complanatum*, 53
Abiotic effects, 10–12. *See also* Biotic effects
AdHu5. *See* Human adenovirus serotype 5
AgAPN1. *See A. gambiae* aminopeptidase
Alhydrogel®, 126–127
Aluminium salts, 126–127
Ambuscade, 19
Ambush behaviour. *See* Ambuscade
Amphitrite ornata (*A. ornata*), 44–45
Annelids, 44–45
Antibody titre, 132
Antigen-presenting cells (APC), 128–129
Asia Pacific Malaria Elimination Network (APMEN), 93–94

B

Baculovirus dual expression system (BDES), 123–124
Baculovirus expression vector systems (BEVS), 123–124
Biotic effects, 12–13. *See also* Abiotic effects
Border crossings, 82–83

C

Candidate vaccines, immunogenicity of, 128
  biodegradable polymeric nano or microparticles, 130
  "carrier" effect, 129
  Pfs25–OMPC, 129f
  PpPfs25 to OMPC, 128
Cell-traversal protein for ookinetes and sporozoites (CelTOS), 119
Cercarial
  emergence
    abiotic effects, 10–12

biotic effects, 12–13
patterns of emergence, 9–10
morphology, 9
swimming, 13
ChAd63. *See* Chimpanzee adenovirus 63
Chaetogaster limnaei (*C. limnaei*), 44–45
Chemical cues, 25–26. *See also* Physical cues
  conspecific inducers, 26
  microbial biofilms, 26–27
  prey species, 27
Chemo-stimulants, 16
Chimpanzee adenovirus 63 (ChAd63), 129–130
Circumsporozoite-thrombospondin related protein (CTRP), 118–119
Connecting Organizations for Regional Disease Surveillance (CORDS), 93
Conspecific inducers, 26
Cotylophoron cotylorum (*C. cotylorum*), 9–10
CpG-oligodeoxynucleotide (CpG-ODN), 126–127
Cross-border malaria, 81
  approaches in tackling, 83b
  cross-border initiatives, 94–96
  and development projects, 89
  displacement due to conflict, 89
  epidemiological drivers, 89–90
    forests and deforestation, 90–91
    misalignment of programmatic approaches, 90
    socioeconomic factors, 91–92
  international collaboration, 92–94
  interventions to addressing, 84b
  migration for work opportunities, 86–87
  movement types across borders, 83b
  patterns of movement, 82–83
  strengthening of preventive measures for, 96
  support operational decision making and surveillance response, 96–97
  surveillance-response system, 94–96

Cross-border malaria (*Continued*)
  treatment-seeking behaviour, 88
  VFR, 85
  visiting friends and relatives, 87–88
Crustacean transport hosts, 38. *See also*
    Molluscan transport hosts; Plant
    transport hosts; Vertebrate
    transport hosts
  field studies, 39–40
  location of metacercariae, 38–39
  P. acanthus, 38
CTRP. *See* Circumsporozoite-
    thrombospondin related protein
Cyst
  aggregations, 23
  associations, 23–24
    F. hepatica, 24–25
    free-living metacercariae, 24

D

Defensive mechanisms, 8
Deforestation, cross-border malaria, 90–91
Direct membrane feeding assay (DMFA),
    128–129
Direct skin feed (DSF), 135–136
Dispersive phase, 13
  cercarial swimming, 13
  duration, 16–17
    ambush behaviour during dispersal, 19
    influence of abiotic factors, 17–18
    H. rhigedana, 19–20
  orientation behaviour, 13–14
    responses to colour, 15–16
    responses to light and gravity, 14
    responses to waterbourne chemicals, 16

E

Echinostomes, 53–54
Epidermal growth factor (EGF), 116
Escherichia coli (*E. coli*), 122

F

Fasciola gigantica (*F. gigantica*), 9–10, 15, 33
Fasciola hepatica (*F. hepatica*), 9–12, 29
Fixed metacercariae morphology, 29–30
  notocotylids, 30
  ventral plug, 30–31

Floating metacercariae morphology, 31
Forests, cross-border malaria, 90–91
Free-floating metacercariae, 46
  principal free-floating metacercariae,
    46–47
  secondary free-floating metacercariae,
    48
    additional mechanisms, 49–51
    epidemiological significance, 51–52
    floating cyst production, 49
    laboratory and field experiments, 49
    liver fluke biology, 48–49
    Megalodiscus species, 49
Free-living metacercariae, 3
  aberrant free-living existence, 52–54
  pollution and, 60
    changes in viability of metacercariae,
      60–61
    dispersal phase pollution, 62
    emergence of cercariae, 60
    length of cercarial dispersive period, 60
    tegumental surface and mineral outer
      shell, 61–62
    variable effect of pollutants, 61
    in vivo infectivity studies, 61
  presettlement phase, 9–20
  settlement phase, 20–52
  transport hosts and metacercariae, 4–5, 8
    avoidance of colonization, 7–8
    colonization by parasites, 7
    dangers of physiological stress, 6
    defensive mechanisms, 8
    ecological and physiological
      homeostasis, 7
    fluctuations in physiological activity, 6
    host choice, 5–6
    instability, 6
    substrate, 5

G

Gastroliths, 46
Gizzard/stomach stones. *See* Gastroliths
Global Malaria Action Plan (GMAP2), 81
Glycosylphosphatidylinositol anchor (GPI
    anchor), 115
Gold standard' assay, 132. *See also*
    Laboratory-based population assay

Index

H
Himasthla rhigedana (*H. rhigedana*), 15–16, 19–20
Human adenovirus serotype 5 (AdHu5), 129–130

I
Indoor residual spraying (IRS), 80–81
International collaboration, 92–94
Intramuscular immunization (IM immunization), 123–124
Intranasal immunization (IN immunization), 123–124
IRS. *See* Indoor residual spraying

L
Laboratory feeding assays, mosquito infection in, 132–136
Laboratory-based population assay, 136
 TPP and path forward, 137–138
Lecithotrophic larval swimming speed, 13
Lectins, 120
Long-lasting insecticide-treated bed net (LLIN), 80–81
Lubombo Spatial Development Initiative (LSDI), 92–93

M
Malaria, 80–81, 110. *See also* Cross-border malaria
 elimination requirement, 82b
 GMAP2, 81
 life cycle, 111–112, 111f
Maltose-binding protein (MBP), 124–125. *See also* Ookinete-secreted proteins
MAOP. *See* Membrane-attack, ookinete protein
Marine syncoeliids, 47
MECIDS. *See* Middle East Consortium on Infectious Disease Surveillance
Megalodiscus species, 49
Megalodiscus temperatus (*M. temperatus*), 16
Mekong Basin Disease Surveillance (MBDS), 93
Membrane-attack, ookinete protein (MAOP), 119

Metacercarial biology. *See also* Free-living metacercariae
 metacercarial excystment, 58
  activation, 58
  duration, 59
  first sensory stimulus, 58–59
 metacercarial infectivity, 57–58
 metacercarial viability, 54
  abiotic effects, 54–57
Metacercarial cyst morphology
 fixed metacercariae, 29–30
  Notocotylids, 30
  ventral plug, 30–31
 floating metacercariae, 31
Metacercarial excystment, 58
 activation, 58
 duration, 59
 first sensory stimulus, 58–59
Metacercarial infectivity, 57–58
Metacercarial viability, 54
 abiotic effects, 54–57
Microbial biofilms, 26–27
Middle East Consortium on Infectious Disease Surveillance (MECIDS), 93
Minimum emergence temperature threshold (METT), 10–11
Modified vaccinia virus Ankara (MVA), 129–130
Molluscan transport hosts, 33–34. *See also* Crustacean transport hosts; Plant transport hosts; Vertebrate transport hosts
 biotic factors influencing encystment success, 37
 field studies, 35
  Philophthalmus sp. metacercariae, 35
  running water influence, 36–37
  spatial and temporal variability, 36
  Sphaeridotrema spp. Metacercariae, 35–36
 location of metacercariae, 34–35
Montanide^{TM}, 127
Mosquito
 components, 119–120
  CPBAg1, 120–121
  lectins, 120

Mosquito (*Continued*)
   plasmodium infectivity, 121
   saglin, 121
   infection
      DMFA, 135–136
      DSF, 135–136
      in laboratory feeding assays, 132
      SMFA and DMFA, 133–135
Multicomponent vaccines, 130–131
MVA. *See* Modified vaccinia virus Ankara
Mytilus edulis (*M. edulis*), 37

N

Notocotylids, 9–10, 30, 35
Notocotylus attenuatus (*N. attenuatus*), 32–33
Notocotylus ralli (*N. ralli*), 14

O

Ookinete-secreted proteins, 117–118
   ookinete micronemal proteins, 118
   PPLPs, 119
   SOAP, 119
Orientation behaviour, 13–14
   responses to colour, 15–16
   responses to light and gravity, 14
   responses to waterbourne chemicals, 16
Outer membrane protein complex (OMPC), 128

P

Paramonostomum chabaudi (*P. chabaudi*), 32
Parasite surface
   sequence polymorphism, 116–117
   structural features, spatial and temporal expression, 113–114
      anti-Pfs230 antibodies, 114–115
      P230, 114
      Pfs25 and Pfs28, 116
      Pfs48/45 protein, 115
      Plasmodium spp., 115
Parorchis acanthus (*P. acanthus*), 10–12, 23–24
Peritrophic matrix (PM), 110–111, 118
Philophthalmus sp., 12
Physical cues, 27–28. *See also* Chemical cues
   cercarial tactile responses, 28–29
   rheotaxis, 28

rugotropism, 28
   surface textures and contours, 28
Pichia pastoris (*P. pastoris*), 123
Plant transport hosts, 40. *See also* Crustacean transport hosts; Molluscan transport hosts; Vertebrate transport hosts
   ephemeral nature, 40
   field studies, 43–44
   location of metacercariae, 41–42
   marine and freshwater environments, 40–41
   substrate, 40
Plasmodium fallax (*P. fallax*), 112
Plasmodium gallinaceum (*P. gallinaceum*), 112
Plasmodium perforin-like protein (PPLP), 119
PM. *See* Peritrophic matrix
Pollution of aquatic habitats, 60
Positive phototaxis, 14
Post-activation targets, 113–117
Pre-activation targets, 113–117
Presettlement phase, free-living metacercariae. *See also* Settlement phase, free-living metacercariae
   cercarial emergence
      abiotic effects, 10–12
      biotic effects, 12–13
      patterns of emergence, 9–10
   cercarial morphology, 9
   dispersive phase, 13
      cercarial swimming, 13
      duration, 16–19
      orientation behaviour, 13–16
Prey species, 27
Principal free-floating metacercariae, 46–47. *See also* Secondary free-floating metacercariae
Rapid diagnostic test (RDT), 80–81
Recombinant vaccines production, challenge in, 122
   76 kDa segment C, 124–125
   cysteine-rich parasite surface proteins, 126
   P. pastoris expression system, 123
   purification process, 123–124
   recombinant P48/45, 125
   S. cerevisiae, 122–123

Relative humidity (RH), 56
Reproduction number (R), 131–132
Rheotaxis, 28
Rugotropism, 28

S

Saccharomyces cerevisiae (*S. cerevisiae*), 122–123
Saccocoelioides octavus (*S. octavus*), 29
Salinity, 11
SDSS. *See* Spatial decision support system
Secondary free-floating metacercariae, 48.
  See also Principal free-floating
  metacercariae
  additional mechanisms, 49–51
  epidemiological significance, 51–52
  floating cyst production, 49
  laboratory and field experiments, 49
  liver fluke biology, 48–49
  Megalodiscus species, 49
Secreted ookinete adhesive protein
  (SOAP), 119
Sequence polymorphism, 116–117
Settlement phase, free-living
  metacercariae. *See also*
  Presettlement phase, free-living
  metacercariae
  free-floating metacercariae, 46–52
  gregarious and substrate-associated
    settlement behaviour, 21
    cyst aggregations, 23
    cyst associations, 23–25
  metacercarial cyst morphology
    fixed metacercariae, 29–31
    floating metacercariae, 31
  settlement and encystment, 20–21, 22f
  settlement and encystment cues, 25
    chemical cues, 25–27
    for free-floating metacercariae, 29
    physical cues, 27–29
  transport hosts
    Crustacean, 38–40
    mechanisms, 46
    miscellaneous transport hosts, 44–46
    Molluscan, 33–37
    plant, 40–44
    selection, 31–33
"Sinkers", 83b

SOAP. *See* Secreted ookinete adhesive
  protein
Socioeconomic factors, 91–92
Spatial decision support system (SDSS),
  96–97
Sphaeridiotrema globulus metacercariae, 24
Standard membrane feeding assay (SMFA),
  113
Substrate, 5, 40
Surveillance-response system, 94–96

T

Target Product Profile (TPP), 130
  and path forward, 137–138
Temperature, 10–11
Tetanus toxoid (TT), 128
Thigmotaxis, 28–29
Thrombospondin-related adhesive protein
  (TRAP), 118–119
TPP. *See* Target Product Profile
TRA. *See* Transmission-reducing activity
Transmission-blocking activity (TBA), 113
Transmission-blocking vaccine (TBV),
  111–112
  protective correlates and surrogate assays
    for efficacy, 131–132
    antibody titre, 132
    laboratory-based population assay,
      136–138
    mosquito infection in laboratory
      feeding assays, 132–136
  targets for, 113
    mosquito components, 119–121
    ookinete-secreted proteins, 117–119
    parasite surface, 113–117
  vaccine development efforts and status,
    121–122
    challenge in recombinant vaccines
      production, 122–126
    challenges in vaccine formulation,
      126–127
    immunogenicity of candidate vaccines,
      128–130
    multicomponent vaccines, 130–131
  vaccines features, 112–113
Transmission-reducing activity (TRA),
  113

Transmission-reducing antibodies, 112
Transport hosts
   Crustacean, 38–40
   mechanisms, 46
   and metacercariae, 4–5, 8
      avoidance of colonization, 7–8
      colonization by parasites, 7
      dangers of physiological stress, 6
      defensive mechanisms, 8
      ecological and physiological homeostasis, 7
      fluctuations in physiological activity, 6
      host choice, 5–6
      instability, 6
      substrate, 5
   miscellaneous transport hosts, 44–46
   Molluscan, 33–37
   plant, 40–44
   selection, 31–33
TRAP. *See* Thrombospondin-related adhesive protein

Trematode anklets, 47
TT. *See* Tetanus toxoid

U
United Arab Emirates (UAE), 86

V
Vaccine formulation, challenges in, 126–127
Vaccines, 112–113. *See also* Transmission-blocking vaccines (TBV)
Vertebrate transport hosts, 45–46. *See also* Crustacean transport hosts; Molluscan transport hosts; Plant transport hosts
Viral-like particle (VLP), 123
Visiting friends and relative (VFR), 85

W
Willebrand factor-A domain-related protein (WARP), 118
World Health Organization (WHO), 80–81

CONTENTS OF VOLUMES IN THIS SERIES

Volume 41

Drug Resistance in Malaria Parasites of Animals and Man
W. Peters

Molecular Pathobiology and Antigenic Variation of *Pneumocystis carinii*
Y. Nakamura and M. Wada

Ascariasis in China
P. Weidono, Z. Xianmin and D.W.T. Crompton

The Generation and Expression of Immunity to *Trichinella spiralis* in Laboratory Rodents
R.G. Bell

Population Biology of Parasitic Nematodes: Application of Genetic Markers
T.J.C. Anderson, M.S. Blouin and R.M. Brech

Schistosomiasis in Cattle
J. De Bont and J. Vercruysse

Volume 42

The Southern Cone Initiative Against Chagas Disease
C. J. Schofield and J.C.P. Dias

Phytomonas and Other Trypanosomatid Parasites of Plants and Fruit
E.P. Camargo

Paragonimiasis and the Genus *Paragonimus*
D. Blair, Z.-B. Xu, and T. Agatsuma

Immunology and Biochemistry of *Hymenolepis diminuta*
J. Anreassen, E.M. Bennet-Jenkins, and C. Bryant

Control Strategies for Human Intestinal Nematode Infections
M. Albonico, D.W.T. Crompton, and L. Savioli

DNA Vaocines: Technology and Applications as Anti-parasite and Anti-microbial Agents
J.B. Alarcon, G.W. Wainem and D.P. McManus

Volume 43

Genetic Exchange in the Trypanosomatidae
W. Gibson and J. Stevens

The Host-Parasite Relationship in Neosporosis
A. Hemphill

Proteases of Protozoan Parasites
P.J. Rosenthal

Proteinases and Associated Genes of Parasitic Helminths
J. Tort, P.J. Brindley, D. Knox, K.H. Wolfe, and J.P. Dalton

Parasitic Fungi and their Interaction with the Insect Immune System
A. Vilcinskas and P. Götz

Volume 44

Cell Biology of *Leishmania*
B. Handman

Immunity and Vaccine Development in the Bovine Theilerioses
N. Boulter and R. Hall

The Distribution of *Schistosoma bovis* Sonaino, 1876 in Relation to Intermediate Host Mollusc-Parasite Relationships
H. Moné, G. Mouahid, and S. Morand

The Larvae of Monogenea (Platyhelminthes)
H.D. Whittington, L.A. Chisholm, and K. Rohde

Sealice on Salmonids: Their Biology and Control
A.W. Pike and S.L. Wadsworth

Volume 45

The Biology of some Intraerythrocytic Parasites of Fishes, Amphibia and Reptiles
A.J. Davies and M.R.L. Johnston

The Range and Biological Activity of FMR Famide-related Peptides and Classical Neurotransmitters in Nematodes
D. Brownlee, L. Holden-Dye, and R. Walker

The Immunobiology of Gastrointestinal
Nematode Infections in Ruminants
A. Balic, V.M. Bowles, and E.N.T. Meeusen

Volume 46

Host-Parasite Interactions in Acanthocephala:
A Morphological Approach
H. Taraschewski

Eicosanoids in Parasites and Parasitic Infections
A. Daugschies and A. Joachim

Volume 47

An Overview of Remote Sensing and Geodesy
for Epidemiology and Public Health
Application
S.I. Hay

Linking Remote Sensing, Land Cover and
Disease
*P.J. Curran, P.M. Atkinson, G.M. Foody, and
E.J. Milton*

Spatial Statistics and Geographic Information
Systems in Epidemiology and Public
Health
T.P. Robinson

Satellites, Space, Time and the African
Trypanosomiases
D.J. Rogers

Earth Observation, Geographic Information
Systems and *Plasmodium falciparum* Malaria
in Sub-Saharan Africa
*S.I. Hay, J. Omumbo, M. Craig, and
R.W. Snow*

Ticks and Tick-borne Disease Systems in Space
and from Space
S.E. Randolph

The Potential of Geographical Information
Systems (GIS) and Remote Sensing in the
Epidemiology and Control of Human
Helminth Infections
S. Brooker and E. Michael

Advances in Satellite Remote Sensing
of Environmental Variables for
Epidemiological Applications
S.J. Goetz, S.D. Prince, and J. Small

Forecasting Diseases Risk for Increased
Epidemic Preparedness in Public Health
*M.F. Myers, D.J. Rogers, J. Cox, A. Flauhalt,
and S.I. Hay*

Education, Outreach and the Future of
Remote Sensing in Human Health
*B.L. Woods, L.R. Beck, B.M. Lobitz, and
M.R. Bobo*

Volume 48

The Molecular Evolution of
Trypanosomatidae
*J.R. Stevens, H.A. Noyes, C.J. Schofield, and
W. Gibson*

Transovarial Transmission in the Microsporidia
A.M. Dunn, R.S. Terry, and J.E. Smith

Adhesive Secretions in the Platyhelminthes
I.D. Whittington and B.W. Cribb

The Use of Ultrasound in Schistosomiasis
C.F.R. Hatz

Ascaris and Ascariasis
D.W.T. Crompton

Volume 49

Antigenic Variation in Trypanosomes:
Enhanced Phenotypic Variation in a
Eukaryotic Parasite
H.D. Barry and R. McCulloch

The Epidemiology and Control of Human
African Trypanosomiasis
J. Pépin and H.A. Méda

Apoptosis and Parasitism: from the Parasite to
the Host Immune Response
G.A. DosReis and M.A. Barcinski

Biology of Echinostomes Except *Echinostoma*
B. Fried

Volume 50

The Malaria-Infected Red Blood Cell:
Structural and Functional Changes
B.M. Cooke, N. Mohandas, and R.L. Coppel

Schistosomiasis in the Mekong Region:
Epidemiology and Phytogeography
S.W. Attwood

Molecular Aspects of Sexual Development
and Reproduction in Nematodes and
Schistosomes
P.R. Boag, S.E. Newton, and R.B. Gasser

Antiparasitic Properties of Medicinal Plants and
Other Naturally Occurring Products
S. Tagboto and S. Townson

Volume 51

Aspects of Human Parasites in which Surgical
Intervention May Be Important
D.A. Meyer and B. Fried

Electron-transfer Complexes in *Ascaris*
Mitochondria
K. Kita and S. Takamiya

Cestode Parasites: Application of *In Vivo* and
In Vitro Models for Studies of the
Host-Parasite Relationship
M. Siles-Lucas and A. Hemphill

Volume 52

The Ecology of Fish Parasites with Particular
Reference to Helminth Parasites and their
Salmonid Fish Hosts in Welsh Rivers:
A Review of Some of the Central
Questions
J.D. Thomas

Biology of the Schistosome Genus
Trichobilharzia
P. Horak, L. Kolarova, and C.M. Adema

The Consequences of Reducing Transmission
of *Plasmodium falciparum* in Africa
R.W. Snow and K. Marsh

Cytokine-Mediated Host Responses during
Schistosome Infections: Walking the Fine
Line Between Immunological Control
and Immunopathology
K.F. Hoffmann, T.A. Wynn, and
D.W. Dunne

Volume 53

Interactions between Tsetse and
Trypanosomes with Implications for the
Control of Trypanosomiasis
S. Aksoy, W.C. Gibson, and M.J. Lehane

Enzymes Involved in the Biogenesis of the
Nematode Cuticle
A.P. Page and A.D. Winter

Diagnosis of Human Filariases (Except
Onchocerciasis)
M. Walther and R. Muller

Volume 54

Introduction — Phylogenies, Phylogenetics,
Parasites and the Evolution of Parasitism
D.T.J. Littlewood

Cryptic Organelles in Parasitic Protists and Fungi
B.A.P. Williams and P.J. Keeling

Phylogenetic Insights into the Evolution
of Parasitism in Hymenoptera
J.B. Whitfield

Nematoda: Genes, Genomes and the
Evolution of Parasitism
M.L. Blaxter

Life Cycle Evolution in the Digenea: A New
Perspective from Phylogeny
T.H. Cribb, R.A. Bray, P.D. Olson, and
D.T. J. Littlewood

Progress in Malaria Research: The Case for
Phylogenetics
S.M. Rich and F.J. Ayala

Phylogenies, the Comparative Method and
Parasite Evolutionary Ecology
S. Morand and R. Poulin

Recent Results in Cophylogeny Mapping
M.A. Charleston

Inference of Viral Evolutionary Rates from
Molecular Sequences
A. Drummond, O.G. Pybus, and A. Rambaut

Detecting Adaptive Molecular Evolution:
Additional Tools for the Parasitologist
J.O. McInerney, D.T.J. Littlewood, and
C.J. Creevey

Volume 55

Contents of Volumes 28–52
Cumulative Subject Indexes for Volumes
28–52
Contributors to Volumes 28–52

Volume 56

Glycoinositolphospholipid from *Trypanosoma cruzi*: Structure, Biosynthesis and Immunobiology
J.O. Previato, R. Wait, C. Jones, G.A. DosReis, A.R. Todeschini, N. Heise and L.M. Previata

Biodiversity and Evolution of the Myxozoa
E.U. Canning and B. Okamura

The Mitochondrial Genomics of Parasitic Nematodes of Socio-Economic Importance: Recent Progress, and Implications for Population Genetics and Systematics
M. Hu, N.B. Chilton, and R.B. Gasser

The Cytoskeleton and Motility in Apicomplexan Invasion
R.E. Fowler, G. Margos, and G.H. Mitchell

Volume 57

Canine Leishmaniasis
J. Alvar, C. Cañavate, R. Molina, J. Moreno, and J. Nieto

Sexual Biology of Schistosomes
H. Moné and J. Boissier

Review of the Trematode Genus *Ribeiroia* (Psilostomidae): Ecology, Life History, and Pathogenesis with Special Emphasis on the Amphibian Malformation Problem
P.T.J. Johnson, D.R. Sutherland, J.M. Kinsella and K.B. Lunde

The *Trichuris muris* System: A Paradigm of Resistance and Susceptibility to Intestinal Nematode Infection
L.J. Cliffe and R.K. Grencis

Scabies: New Future for a Neglected Disease
S.F. Walton, D.C. Holt, B.J. Currie, and D.J. Kemp

Volume 58

Leishmania spp.: On the Interactions they Establish with Antigen-Presenting Cells of their Mammalian Hosts
J.-C. Antoine, E. Prina, N. Courret, and T. Lang

Variation in *Giardia*: Implications for Taxonomy and Epidemiology
R.C.A. Thompson and P.T. Monis

Recent Advances in the Biology of *Echinostoma* species in the "revolutum" Group
B. Fried and T.K. Graczyk

Human Hookworm Infection in the 21st Century
S. Brooker, J. Bethony, and P.J. Hotez

The Curious Life-Style of the Parasitic Stages of Gnathiid Isopods
N.J. Smit and A.J. Davies

Volume 59

Genes and Susceptibility to Leishmaniasis
Emanuela Handman, Colleen Elso, and Simon Foote

Cryptosporidium and Cryptosporidiosis
R.C.A. Thompson, M.E. Olson, G. Zhu, S. Enomoto, Mitchell S. Abrahamsen and N.S. Hijjawi

Ichthyophthirius multifiliis Fouquet and Ichthyophthiriosis in Freshwater Teleosts
R.A. Matthews

Biology of the Phylum Nematomorpha
B. Hanelt, F. Thomas, and A. Schmidt-Rhaesa

Volume 60

Sulfur-Containing Amino Acid Metabolism in Parasitic Protozoa
Tomoyoshi Nozaki, Vahab Ali, and Masaharu Tokoro

The Use and Implications of Ribosomal DNA Sequencing for the Discrimination of Digenean Species
Matthew J. Nolan and Thomas H. Cribb

Advances and Trends in the Molecular Systematics of the Parasitic Platyhelminthes
Peter D. Olson and Vasyl V. Tkach

Wolbachia Bacterial Endosymbionts of Filarial Nematodes
Mark J. Taylor, Claudio Bandi, and Achim Hoerauf

The Biology of Avian *Eimeria* with an Emphasis on their Control by Vaccination
Martin W. Shirley, Adrian L. Smith, and Fiona M. Tomley

Volume 61

Control of Human Parasitic Diseases: Context and Overview
David H. Molyneux

Malaria Chemotherapy
Peter Winstanley and Stephen Ward

Insecticide-Treated Nets
Jenny Hill, Jo Lines, and Mark Rowland

Control of Chagas Disease
Yoichi Yamagata and Jun Nakagawa

Human African Trypanosomiasis: Epidemiology and Control
E.M. Févre, K. Picozzi, J. Jannin, S.C. Welburn and I. Maudlin

Chemotherapy in the Treatment and Control of Leishmaniasis
Jorge Alvar, Simon Croft, and Piero Olliaro

Dracunculiasis (Guinea Worm Disease) Eradication
Ernesto Ruiz-Tiben and Donald R. Hopkins

Intervention for the Control of Soil-Transmitted Helminthiasis in the Community
Marco Albonico, Antonio Montresor, D.W.T. Crompton, and Lorenzo Savioli

Control of Onchocerciasis
Boakye A. Boatin and Frank O. Richards, Jr.

Lymphatic Filariasis: Treatment, Control and Elimination
Eric A. Ottesen

Control of Cystic Echinococcosis/Hydatidosis: 1863-2002
P.S. Craig and E. Larrieu

Control of *Taenia solium* Cysticercosis/Taeniosis
Arve Lee Willingham III and Dirk Engels

Implementation of Human Schistosomiasis Control: Challenges and Prospects
Alan Fenwick, David Rollinson, and Vaughan Southgate

Volume 62

Models for Vectors and Vector-Borne Diseases
D.J. Rogers

Global Environmental Data for Mapping Infectious Disease Distribution
S.I. Hay, A.J. Tatem, A.J. Graham, S.J. Goetz, and D.J. Rogers

Issues of Scale and Uncertainty in the Global Remote Sensing of Disease
P.M. Atkinson and A.J. Graham

Determining Global Population Distribution: Methods, Applications and Data
D.L. Balk, U. Deichmann, G. Yetman, F. Pozzi, S.I. Hay, and A. Nelson

Defining the Global Spatial Limits of Malaria Transmission in 2005
C.A. Guerra, R.W. Snow and S.I. Hay

The Global Distribution of Yellow Fever and Dengue
D.J. Rogers, A.J. Wilson, S.I. Hay, and A.J. Graham

Global Epidemiology, Ecology and Control of Soil-Transmitted Helminth Infections
S. Brooker, A.C.A. Clements and D.A.P. Bundy

Tick-borne Disease Systems: Mapping Geographic and Phylogenetic Space
S.E. Randolph and D.J. Rogers

Global Transport Networks and Infectious Disease Spread
A.J. Tatem, D.J. Rogers and S.I. Hay

Climate Change and Vector-Borne Diseases
D.J. Rogers and S.E. Randolph

Volume 63

Phylogenetic Analyses of Parasites in the New Millennium
David A. Morrison

Targeting of Toxic Compounds to the
  Trypanosome's Interior
  Michael P. Barrett and Ian H. Gilbert

Making Sense of the Schistosome Surface
  Patrick J. Skelly and R. Alan Wilson

Immunology and Pathology of Intestinal
  Trematodes in Their Definitive Hosts
  *Rafael Toledo, José-Guillermo Esteban, and
  Bernard Fried*

Systematics and Epidemiology of *Trichinella*
  Edoardo Pozio and K. Darwin Murrell

Volume 64

Leishmania and the Leishmaniases: A Parasite
  Genetic Update and Advances in
  Taxonomy, Epidemiology and
  Pathogenicity in Humans
  *Anne-Laure Bañuls, Mallorie Hide and
  Franck Prugnolle*

Human Waterborne Trematode and
  Protozoan Infections
  Thaddeus K. Graczyk and Bernard Fried

The Biology of Gyrodctylid Monogeneans:
  The "Russian-Doll Killers"
  T.A. Bakke, J. Cable, and P.D. Harris

Human Genetic Diversity and the
  Epidemiology of Parasitic and Other
  Transmissible Diseases
  Michel Tibayrenc

Volume 65

ABO Blood Group Phenotypes and
  Plasmodium falciparum Malaria: Unlocking
  a Pivotal Mechanism
  *María-Paz Loscertales, Stephen Owens,
  James O'Donnell, James Bunn,
  Xavier Bosch-Capblanch, and
  Bernard J. Brabin*

Structure and Content of the Entamoeba
  histolytica Genome
  *C.G. Clark, U.C.M. Alsmark,
  M. Tazreiter, Y. Saito-Nakano, V. Ali,
  S. Marion, C. Weber, C. Mukherjee,
  I. Bruchhaus, E. Tannich, M. Leippe,*
  *T. Sicheritz-Ponten, P. G. Foster,
  J. Samuelson, C.J. Noël, R.P. Hirt,
  T.M. Embley, C. A. Gilchrist,
  B.J. Mann, U. Singh, J.P. Ackers,
  S. Bhattacharya, A. Bhattacharya,
  A. Lohia, N. Guillén, M. Duchene,
  T. Nozaki, and N. Hall*

Epidemiological Modelling for Monitoring
  and Evaluation of Lymphatic Filariasis
  Control
  *Edwin Michael, Mwele N. Malecela-Lazaro,
  and James W. Kazura*

The Role of Helminth Infections in
  Carcinogenesis
  David A. Mayer and Bernard Fried

A Review of the Biology of the
  Parasitic Copepod Lernaeocera
  branchialis (L., 1767)(Copepoda:
  Pennellidae
  *Adam J. Brooker, Andrew P. Shinn, and
  James E. Bron*

Volume 66

Strain Theory of Malaria: The First 50 Years
  F. Ellis McKenzie, David L. Smith,
  Wendy P. O'Meara, and
  Eleanor M. Riley*

Advances and Trends in the Molecular
  Systematics of Anisakid Nematodes, with
  Implications for their Evolutionary
  Ecology and Host—Parasite
  Co-evolutionary Processes
  Simonetta Mattiucci and Giuseppe Nascetti

Atopic Disorders and Parasitic Infections
  Aditya Reddy and Bernard Fried

Heartworm Disease in Animals and Humans
  *John W. McCall, Claudio Genchi, Laura
  H. Kramer, Jorge Guerrero, and
  Luigi Venco*

Volume 67

Introduction
  Irwin W. Sherman

An Introduction to Malaria Parasites
  Irwin. W. Sherman

The Early Years
Irwin W. Sherman

Show Me the Money
Irwin W. Sherman

In Vivo and In Vitro Models
Irwin W. Sherman

Malaria Pigment
Irwin W. Sherman

Chloroquine and Hemozoin
Irwin W. Sherman

Invasion of Erythrocytes
Irwin W. Sherman

Vitamins and Anti-Oxidant Defenses
Irwin W. Sherman

Shocks and Clocks
Irwin W. Sherman

Transcriptomes, Proteomes and Data Mining
Irwin W. Sherman

Mosquito Interactions
Irwin W. Sherman

Isoenzymes
Irwin W. Sherman

The Road to the Plasmodium falciparum Genome
Irwin W. Sherman

Carbohydrate Metabolism
Irwin W. Sherman

Pyrimidines and the Mitochondrion
Irwin W. Sherman

The Road to Atovaquone
Irwin W. Sherman

The Ring Road to the Apicoplast
Irwin W. Sherman

Ribosomes and Ribosomal Ribonucleic Acid Synthesis
Irwin W. Sherman

De Novo Synthesis of Pyrimidines and Folates
Irwin W. Sherman

Salvage of Purines
Irwin W. Sherman

Polyamines
Irwin W. Sherman

New Permeability Pathways and Transport
Irwin W. Sherman

Hemoglobinases
Irwin W. Sherman

Erythrocyte Surface Membrane Proteins
Irwin W. Sherman

Trafficking
Irwin W. Sherman

Erythrocyte Membrane Lipids
Irwin W. Sherman

Volume 68

HLA-Mediated Control of HIV and HIV Adaptation to HLA
Rebecca P. Payne, Philippa C. Matthews, Julia G. Prado, and Philip J.R. Goulder

An Evolutionary Perspective on Parasitism as a Cause of Cancer
Paul W. Ewald

Invasion of the Body Snatchers: The Diversity and Evolution of Manipulative Strategies in Host–Parasite Interactions
Thierry Lefèvre, Shelley A. Adamo, David G. Biron, Dorothée Missé, David Hughes, and Frédéric Thomas

Evolutionary Drivers of Parasite-Induced Changes in Insect Life-History Traits: From Theory to Underlying Mechanisms
Hilary Hurd

Ecological Immunology of a Tapeworms' Interaction with its Two Consecutive Hosts
Katrin Hammerschmidt and Joachim Kurtz

Tracking Transmission of the Zoonosis Toxoplasma gondii
Judith E. Smith

Parasites and Biological Invasions
Alison M. Dunn

Zoonoses in Wildlife: Integrating Ecology into Management
Fiona Mathews

Understanding the Interaction Between an Obligate Hyperparasitic Bacterium, Pasteuria penetrans and its Obligate Plant-Parasitic Nematode Host, Meloidogyne spp.
Keith G. Davies

Host—Parasite Relations and Implications for Control
Alan Fenwick

Onchocerca—Simulium Interactions and the Population and Evolutionary Biology of Onchocerca volvulus
María-Gloria Basáñez, Thomas S. Churcher, and María-Eugenia Grillet

Microsporidians as Evolution-Proof Agents of Malaria Control?
Jacob C. Koella, Lena Lorenz, and Irka Bargielowski

Volume 69

The Biology of the Caecal Trematode Zygocotyle lunata
Bernard Fried, Jane E. Huffman, Shamus Keeler, and Robert C. Peoples

Fasciola, Lymnaeids and Human Fascioliasis, with a Global Overview on Disease Transmission, Epidemiology, Evolutionary Genetics, Molecular Epidemiology and Control
Santiago Mas-Coma, María Adela Valero, and María Dolores Bargues

Recent Advances in the Biology of Echinostomes
Rafael Toledo, José-Guillermo Esteban, and Bernard Fried

Peptidases of Trematodes
Martin Kašný, Libor Mikeš, Vladimír Hampl, Jan Dvořák, Conor R. Caffrey, John P. Dalton, and Petr Horák

Potential Contribution of Sero-Epidemiological Analysis for Monitoring Malaria Control and Elimination: Historical and Current Perspectives
Chris Drakeley and Jackie Cook

Volume 70

Ecology and Life History Evolution of Frugivorous Drosophila Parasitoids
Frédéric Fleury, Patricia Gibert, Nicolas Ris, and Roland Allemand

Decision-Making Dynamics in Parasitoids of Drosophil
Andra Thiel and Thomas S. Hoffmeister

Dynamic Use of Fruit Odours to Locate Host Larvae: Individual Learning, Physiological State and Genetic Variability as Adaptive Mechanisms
Laure Kaiser, Aude Couty, and Raquel Perez-Maluf

The Role of Melanization and Cytotoxic By-Products in the Cellular Immune Responses of Drosophila Against Parasitic Wasps
A. Nappi, M. Poirié, and Y. Carton

Virulence Factors and Strategies of Leptopilina spp.: Selective Responses in Drosophila Hosts
Mark J. Lee, Marta E. Kalamarz, Indira Paddibhatla, Chiyedza Small, Roma Rajwani, and Shubha Govind

Variation of Leptopilina boulardi Success in Drosophila Hosts: What is Inside the Black Box?
A. Dubuffet, D. Colinet, C. Anselme, S. Dupas, Y. Carton, and M. Poirié

Immune Resistance of Drosophila Hosts Against Asobara Parasitoids: Cellular Aspects
Patrice Eslin, Geneviève Prévost, Sebastien Havard, and Géraldine Doury

Components of Asobara Venoms and their Effects on Hosts
Sébastien J.M. Moreau, Sophie Vinchon, Anas Cherqui, and Geneviève Prévost

Strategies of Avoidance of Host Immune Defenses in Asobara Species
Geneviève Prevost, Géraldine Doury, Alix D.N. Mabiala-Moundoungou, Anas Cherqui, and Patrice Eslin

Evolution of Host Resistance and Parasitoid Counter-Resistance
Alex R. Kraaijeveld and H. Charles J. Godfray

Local, Geographic and Phylogenetic Scales of Coevolution in *Drosophila*—Parasitoid Interactions
S. Dupas, A. Dubuffet, Y. Carton, and M. Poirié

Drosophila—Parasitoid Communities as Model Systems for Host—*Wolbachia* Interactions
Fabrice Vavre, Laurence Mouton, and Bart A. Pannebakker

A Virus-Shaping Reproductive Strategy in a Drosophila Parasitoid
Julien Varaldi, Sabine Patot, Maxime Nardin, and Sylvain Gandon

Volume 71

Cryptosporidiosis in Southeast Asia: What's out There?
Yvonne A.L. Lim, Aaron R. Jex, Huw V. Smith, and Robin B. Gasser

Human Schistosomiasis in the Economic Community of West African States: Epidemiology and Control
Hélené Moné, Moudachirou Ibikounlé, Achille Massougbodji, and Gabriel Mouahid

The Rise and Fall of Human Oesophagostomiasis
A.M. Polderman, M. Eberhard, S. Baeta, Robin B. Gasser, L. van Lieshout, P. Magnussen, A. Olsen, N. Spannbrucker, J. Ziem, and J. Horton

Volume 72

Important Helminth Infections in Southeast Asia: Diversity, Potential for Control and Prospects for Elimination
Jürg Utzinger, Robert Bergquist, Remigio Olveda, and Xiao-Nong Zhou

Escalating the Global Fight Against Neglected Tropical Diseases Through Interventions in the Asia Pacific Region
Peter J. Hotez and John P. Ehrenberg

Coordinating Research on Neglected Parasitic Diseases in Southeast Asia Through Networking
Remi Olveda, Lydia Leonardo, Feng Zheng, Banchob Sripa, Robert Bergquist, and Xiao-Nong Zhou

Neglected Diseases and Ethnic Minorities in the Western Pacific Region: Exploring the Links
Alexander Schratz, Martha Fernanda Pineda, Liberty G. Reforma, Nicole M. Fox, Tuan Le Anh, L. Tommaso Cavalli-Sforza, Mackenzie K. Henderson, Raymond Mendoza, Jürg Utzinger, John P. Ehrenberg, and Ah Sian Tee

Controlling Schistosomiasis in Southeast Asia: A Tale of Two Countries
Robert Bergquist and Marcel Tanner

Schistosomiasis Japonica: Control and Research Needs
Xiao-Nong Zhou, Robert Bergquist, Lydia Leonardo, Guo-Jing Yang, Kun Yang, M. Sudomo, and Remigio Olveda

Schistosoma mekongi in Cambodia and Lao People's Democratic Republic
Sinuon Muth, Somphou Sayasone, Sophie Odermatt-Biays, Samlane Phompida, Socheat Duong, and Peter Odermatt

Elimination of Lymphatic Filariasis in Southeast Asia
Mohammad Sudomo, Sombat Chayabejara, Duong Socheat, Leda Hernandez, Wei-Ping Wu, and Robert Bergquist

Combating Taenia solium Cysticercosis in Southeast Asia: An Opportunity for Improving Human Health and Livestock Production Links
A. Lee Willingham III, Hai-Wei Wu, James Conlan, and Fadjar Satrija

Echinococcosis with Particular Reference to Southeast Asia
Donald P. McManus

Food-Borne Trematodiases in Southeast Asia: Epidemiology, Pathology, Clinical Manifestation and Control
Banchob Sripa, Sasithorn Kaewkes, Pewpan M. Intapan, Wanchai Maleewong, and Paul J. Brindley

Helminth Infections of the Central Nervous System Occurring in Southeast Asia and the Far East
Shan Lv, Yi Zhang, Peter Steinmann, Xiao-Nong Zhou, and Jürg Utzinger

Less Common Parasitic Infections in Southeast Asia that can Produce Outbreaks
Peter Odermatt, Shan Lv, and Somphou Sayasone

Volume 73

Concepts in Research Capabilities Strengthening: Positive Experiences of Network Approaches by TDR in the People's Republic of China and Eastern Asia
Xiao-Nong Zhou, Steven Wayling, and Robert Bergquist

Multiparasitism: A Neglected Reality on Global, Regional and Local Scale
Peter Steinmann, Jürg Utzinger, Zun-Wei Du, and Xiao-Nong Zhou

Health Metrics for Helminthic Infections
Charles H. King

Implementing a Geospatial Health Data Infrastructure for Control of Asian Schistosomiasis in the People's Republic of China and the Philippines
John B. Malone, Guo-Jing Yang, Lydia Leonardo, and Xiao-Nong Zhou

The Regional Network for Asian Schistosomiasis and Other Helminth Zoonoses (RNAS[+]): Target Diseases in Face of Climate Change
Guo-Jing Yang, Jürg Utzinger, Shan Lv, Ying-Jun Qian, Shi-Zhu Li, Qiang Wang, Robert Bergquist, Penelope Vounatsou, Wei Li, Kun Yang, and Xiao-Nong Zhou

Social Science Implications for Control of Helminth Infections in Southeast Asia
Lisa M. Vandemark, Tie-Wu Jia, and Xiao-Nong Zhou

Towards Improved Diagnosis of Zoonotic Trematode Infections in Southeast Asia
Maria Vang Johansen, Paiboon Sithithaworn, Robert Bergquist, and Jürg Utzinger

The Drugs We Have and the Drugs We Need Against Major Helminth Infections
Jennifer Keiser and Jürg Utzinger

Research and Development of Antischistosomal Drugs in the People's Republic of China: A 60-Year Review
Shu-Hua Xiao, Jennifer Keiser, Ming-Gang Chen, Marcel Tanner, and Jürg Utzinger

Control of Important Helminthic Infections: Vaccine Development as Part of the Solution
Robert Bergquist and Sara Lustigman

Our Wormy World: Genomics, Proteomics and Transcriptomics in East and Southeast Asia
Jun Chuan, Zheng Feng, Paul J. Brindley, Donald P. McManus, Zeguang Han, Peng Jianxin, and Wei Hu

Advances in Metabolic Profiling of Experimental Nematode and Trematode Infections
Yulan Wang, Jia V. Li, Jasmina Saric, Jennifer Keiser, Junfang Wu, Jürg Utzinger, and Elaine Holmes

Studies on the Parasitology, Phylogeography and the Evolution of Host–Parasite Interactions for the Snail Intermediate Hosts of Medically Important Trematode Genera in Southeast Asia
Stephen W. Attwood

Volume 74

The Many Roads to Parasitism: A Tale of Convergence
Robert Poulin

Malaria Distribution, Prevalence, Drug Resistance and Control in Indonesia
Iqbal R.F. Elyazar, Simon I. Hay, and J. Kevin Baird

Cytogenetics and Chromosomes of
   Tapeworms (Platyhelminthes, Cestoda)
   *Marta Špakulová, Martina Orosová, and
   John S. Mackiewicz*

Soil-Transmitted Helminths of Humans in
   Southeast Asia—Towards Integrated
   Control
   *Aaron R. Jex, Yvonne A.L. Lim, Jeffrey
   Bethony, Peter J. Hotez,
   Neil D. Young, and Robin B. Gasser*

The Applications of Model-Based Geostatistics
   in Helminth Epidemiology and Control
   *Ricardo J. Soares Magalhães, Archie C.A.
   Clements, Anand P. Patil,
   Peter W. Gething, and Simon Brooker*

Volume 75

Epidemiology of American Trypanosomiasis
   (Chagas Disease)
   Louis V. Kirchhoff

Acute and Congenital Chagas Disease
   *Caryn Bern, Diana L. Martin, and
   Robert H. Gilman*

Cell-Based Therapy in Chagas Disease
   *Antonio C. Campos de Carvalho,
   Adriana B. Carvalho, and Regina
   C.S. Goldenberg*

Targeting *Trypanosoma cruzi* Sterol
   14α-Demethylase (CYP51)
   *Galina I. Lepesheva, Fernando Villalta,
   and Michael R. Waterman*

Experimental Chemotherapy and Approaches
   to Drug Discovery for *Trypanosoma cruzi*
   Infection
   Frederick S. Buckner

Vaccine Development Against *Trypanosoma
   cruzi* and Chagas Disease
   *Juan C. Vázquez-Chagoyán,
   Shivali Gupta, and Nisha Jain Garg*

Genetic Epidemiology of Chagas Disease
   *Sarah Williams-Blangero, John L. VandeBerg,
   John Blangero, and Rodrigo Corrêa-Oliveira*

Kissing Bugs. The Vectors of Chagas
   *Lori Stevens, Patricia L. Dorn, Justin O. Schmidt,
   John H. Klotz, David Lucero, and
   Stephen A. Klotz*

Advances in Imaging of Animal Models of
   Chagas Disease
   Linda A. Jelicks and Herbert B. Tanowitz

The Genome and Its Implications
   *Santuza M. Teixeira, Najib M. El-Sayed,
   and Patrícia R. Araújo*

Genetic Techniques in *Trypanosoma cruzi*
   *Martin C. Taylor, Huan Huang, and
   John M. Kelly*

Nuclear Structure of *Trypanosoma cruzi*
   *Sergio Schenkman, Bruno dos Santos
   Pascoalino, and Sheila C. Nardelli*

Aspects of Trypanosoma cruzi Stage
   Differentiation
   *Samuel Goldenberg and Andrea
   Rodrigues Ávila*

The Role of Acidocalcisomes in the Stress
   Response of *Trypanosoma cruzi*
   *Roberto Docampo, Veronica Jimenez,
   Sharon King-Keller, Zhu-hong Li, and
   Silvia N.J. Moreno*

Signal Transduction in *Trypanosoma cruzi*
   Huan Huang

Volume 76

Bioactive Lipids in *Trypanosoma cruzi* Infection
   *Fabiana S. Machado, Shankar Mukherjee,
   Louis M. Weiss, Herbert B. Tanowitz, and
   Anthony W. Ashton*

Mechanisms of Host Cell Invasion by
   Trypanosoma cruzi
   *Kacey L. Caradonna and Barbara
   A. Burleigh*

Gap Junctions and Chagas Disease
   *Daniel Adesse, Regina Coeli Goldenberg, Fabio
   S. Fortes, Jasmin, Dumitru A. Iacobas, Sanda
   Iacobas, Antonio Carlos Campos de
   Carvalho, Maria de Narareth
   Meirelles, Huan Huang, Milena B. Soares,
   Herbert B. Tanowitz, Luciana Ribeiro
   Garzoni, and David C. Spray*

The Vasculature in Chagas Disease
   *Cibele M. Prado, Linda A. Jelicks,
   Louis M. Weiss, Stephen M. Factor,
   Herbert B. Tanowitz, and
   Marcos A. Rossi*

Infection-Associated Vasculopathy in
  Experimental Chagas Disease:
  Pathogenic Roles of Endothelin and
  Kinin Pathways
  Julio Scharfstein and Daniele Andrade

Autoimmunity
  *Edecio Cunha-Neto, Priscila Camillo
  Teixeira, Luciana Gabriel Nogueira,
  and Jorge Kalil*

ROS Signalling of Inflammatory Cytokines
  During *Trypanosoma cruzi* Infection
  *Shivali Gupta, Monisha Dhiman, Jian-jun
  Wen, and Nisha Jain Garg*

Inflammation and Chagas Disease: Some
  Mechanisms and Relevance
  André Talvani and Mauro M. Teixeira

Neurodegeneration and Neuroregeneration in
  Chagas Disease
  *Marina V. Chuenkova and Mercio
  PereiraPerrin*

Adipose Tissue, Diabetes and Chagas Disease
  *Herbert B. Tanowitz, Linda A. Jelicks,
  Fabiana S. Machado, Lisia Esper,
  Xiaohua Qi, Mahalia S. Desruisseaux,
  Streamson C. Chua, Philipp E. Scherer,
  and Fnu Nagajyothi*

Volume 77

Coinfection of *Schistosoma* (Trematoda) with
  Bacteria, Protozoa and Helminths
  Amy Abruzzi and Bernard Fried

Trichomonas vaginalis Pathobiology: New
  Insights from the Genome Sequence
  *Robert P. Hirt, Natalia de Miguel,
  Sirintra Nakjang, Daniele Dessi,
  Yuk-Chien Liu, Nicia Diaz,
  Paola Rappelli, Alvaro Acosta-Serrano,
  Pier-Luigi Fiori, and Jeremy C. Mottram*

Cryptic Parasite Revealed: Improved Prospects
  for Treatment and Control of Human
  Cryptosporidiosis Through Advanced
  Technologies
  *Aaron R. Jex, Huw V. Smith, Matthew
  J. Nolan, Bronwyn E. Campbell,
  Neil D. Young, Cinzia Cantacessi, and
  Robin B. Gasser*

Assessment and Monitoring of Onchocerciasis
  in Latin America
  *Mario A. Rodríguez-Pérez,
  Thomas R. Unnasch, and
  Olga Real-Najarro*

Volume 78

Gene Silencing in Parasites: Current Status and
  Future Prospects
  *Raúl Manzano-Román, Ana Oleaga,
  Ricardo Pérez-Sánchez, and
  Mar Siles-Lucas*

Giardia—From Genome to Proteome
  R.C. Andrew Thompson and Paul Monis

Malaria Ecotypes and Stratification
  Allan Schapira and Konstantina Boutsika

The Changing Limits and Incidence of Malaria
  in Africa: 1939–2009
  *Robert W. Snow, Punam Amratia,
  Caroline W. Kabaria, Abdisalan M. Noor,
  and Kevin Marsh*

Volume 79

Northern Host – Parasite Assemblages:
  History and Biogeography on the
  Borderlands of Episodic Climate and
  Environmental Transition
  *Eric P. Hoberg, Kurt E. Galbreath,
  Joseph A. Cook, Susan J. Kutz, and
  Lydden Polley*

Parasites in Ungulates of Arctic North America
  and Greenland: A View of Contemporary
  Diversity, Ecology and Impact in a World
  Under Change
  *Susan J. Kutz, Julie Ducrocq, Guilherme
  G. Verocai, Bryanne M. Hoar, Doug D.
  Colwell, Kimberlee B. Beckmen,
  Lydden Polley, Brett T. Elkin, and
  Eric P. Hoberg*

Neorickettsial Endosymbionts of the Digenea:
  Diversity, Transmission and Distribution
  *Jefferson A. Vaughan, Vasyl V. Tkach, and
  Stephen E. Greiman*

Priorities for the Elimination of Sleeping
  Sickness
  Susan C. Welburn and Ian Maudlin

Scabies: Important Clinical Consequences
  Explained by New Molecular Studies
  *Katja Fischer, Deborah Holt, Bart Currie, and
  David Kemp*

Review: Surveillance of Chagas Disease
  Ken Hashimoto and Kota Yoshioka

Volume 80

The Global Public Health Significance of
  Plasmodium vivax
  *Katherine E. Battle, Peter W. Gething,
  Iqbal R.F. Elyazar,
  Catherine L. Moyes, Marianne E. Sinka,
  Rosalind E. Howes, Carlos A. Guerra,
  Ric N. Price, J. Kevin Baird, and
  Simon I. Hay*

Relapse
  Nicholas J. White and Mallika Imwong

Plasmodium vivax: Clinical Spectrum, Risk
  Factors and Pathogenesis
  *Nicholas M. Anstey, Nicholas M. Douglas,
  Jeanne R. Poespoprodjo, and
  Ric N. Price*

Diagnosis and Treatment of *Plasmodium vivax*
  Malaria
  *J. Kevin Baird, Jason D. Maguire, and
  Ric N. Price*

Chemotherapeutic Strategies for
  Reducing Transmission of *Plasmodium
  vivax* Malaria
  *Nicholas M. Douglas, George K. John,
  Lorenz von Seidlein, Nicholas M. Anstey,
  and Ric N. Price*

Control and Elimination of *Plasmodium vivax*
  G. Dennis Shanks

Volume 81

Plasmodium vivax: Modern Strategies
  to Study a Persistent Parasite's
  Life Cycle
  *Mary R. Galinski, Esmeralda V.S. Meyer, and
  John W. Barnwell*

Red Blood Cell Polymorphism and
  Susceptibility to Plasmodium vivax
  *Peter A. Zimmerman, Marcelo U. Ferreira,
  Rosalind E. Howes, and Odile
  Mercereau-Puijalon*

Natural Acquisition of Immunity to
  Plasmodium vivax: Epidemiological
  Observations and Potential Targets
  *Ivo Mueller, Mary R. Galinski, Takafumi
  Tsuboi, Myriam Arevalo-Herrera,
  William E. Collins, and
  Christopher L. King*

G6PD Deficiency: Global Distribution,
  Genetic Variants and Primaquine Therapy
  *Rosalind E. Howes, Katherine E. Battle,
  Ari W. Satyagraha, J. Kevin Baird, and
  Simon I. Hay*

Genomics, Population Genetics and
  Evolutionary History of *Plasmodium vivax*
  *Jane M. Carlton, Aparup Das, and
  Ananias A. Escalante*

Malariotherapy — Insanity at the Service of
  Malariology
  Georges Snounou and Jean-Louis Pérignon

Volume 82

Recent Developments in Blastocystis Research
  *C. Graham Clark, Mark van der Giezen,
  Mohammed A. Alfellani, and
  C. Rune Stensvold*

Tradition and Transition: Parasitic Zoonoses of
  People and Animals in Alaska, Northern
  Canada, and Greenland
  *Emily J. Jenkins, Louisa J. Castrodale,
  Simone J.C. de Rosemond, Brent R. Dixon,
  Stacey A. Elmore, Karen M. Gesy,
  Eric P. Hoberg, Lydden Polley,
  Janna M. Schurer, Manon Simard, and
  R.C. Andrew Thompson*

The Malaria Transition on the Arabian
  Peninsula: Progress toward a Malaria-Free
  Region between 1960-2010
  *Robert W. Snow, Punam Amratia, Ghasem
  Zamani, Clara W. Mundia, Abdisalan M.
  Noor, Ziad A. Memish, Mohammad H. Al
  Zahrani, Adel AlJasari, Mahmoud Fikri, and
  Hoda Atta*

Microsporidia and 'The Art of Living
  Together'
  Jiří Vávra and Julius Lukeš

Patterns and Processes in Parasite Co-Infection
  Mark E. Viney and Andrea L. Graham

Volume 83

Iron–Sulphur Clusters, Their Biosynthesis, and Biological Functions in Protozoan Parasites
Vahab Ali and Tomoyoshi Nozaki

A Selective Review of Advances in Coccidiosis Research
H. David Chapman, John R. Barta, Damer Blake, Arthur Gruber, Mark Jenkins, Nicholas C. Smith, Xun Suo, and Fiona M. Tomley

The Distribution and Bionomics of *Anopheles* Malaria Vector Mosquitoes in Indonesia
Iqbal R.F. Elyazar, Marianne E. Sinka, Peter W. Gething, Siti N. Tarmidzi, Asik Surya, Rita Kusriastuti, Winarno, J. Kevin Baird, Simon I. Hay, and Michael J. Bangs

Next-Generation Molecular-Diagnostic Tools for Gastrointestinal Nematodes of Livestock, with an Emphasis on Small Ruminants: A Turning Point?
Florian Roeber, Aaron R. Jex, and Robin B. Gasser

Volume 84

Joint Infectious Causation of Human Cancers
Paul W. Ewald and Holly A. Swain Ewald

Neurological and Ocular Fascioliasis in Humans
Santiago Mas-Coma, Verónica H. Agramunt, and María Adela Valero

Measuring Changes in *Plasmodium falciparum* Transmission: Precision, Accuracy and Costs of Metrics
Lucy S. Tusting, Teun Bousema, David L. Smith, and Chris Drakeley

A Review of Molecular Approaches for Investigating Patterns of Coevolution in Marine Host–Parasite Relationships
Götz Froeschke and Sophie von der Heyden

New Insights into Clonality and Panmixia in *Plasmodium* and *Toxoplasma*
Michel Tibayrenc and Francisco J. Ayala

Volume 85

Diversity and Ancestry of Flatworms Infecting Blood of Nontetrapod Craniates "Fishes"
Raphael Orélis-Ribeiro, Cova R. Arias, Kenneth M. Halanych, Thomas H. Cribb, and Stephen A. Bullard

Techniques for the Diagnosis of Fasciola Infections in Animals: Room for Improvement
Cristian A. Alvarez Rojas, Aaron R. Jex, Robin B. Gasser, and Jean-Pierre Y. Scheerlinck

Reevaluating the Evidence for Toxoplasma gondii-Induced Behavioural Changes in Rodents
Amanda R. Worth, R.C. Andrew Thompson, and Alan J. Lymbery

Volume 86

Historical Patterns of Malaria Transmission in China
Jian-Hai Yin, Shui-Sen Zhou, Zhi-Gui Xia, Ru-Bo Wang, Ying-Jun Qian, Wei-Zhong Yang, and Xiao-Nong Zhou

Feasibility and Roadmap Analysis for Malaria Elimination in China
Xiao-Nong Zhou, Zhi-Gui Xia, Ru-Bo Wang, Ying-Jun Qian, Shui-Sen Zhou, Jürg Utzinger, Marcel Tanner, Randall Kramer, and Wei-Zhong Yang

Lessons from Malaria Control to Elimination: Case Study in Hainan and Yunnan Provinces
Zhi-Gui Xia, Li Zhang, Jun Feng, Mei Li, Xin-Yu Feng, Lin-Hua Tang, Shan-Qing Wang, Heng-Lin Yang, Qi Gao, Randall Kramer, Tambo Ernest, Peiling Yap, and Xiao-Nong Zhou

Surveillance and Response to Drive the National Malaria Elimination Programme
Xin-Yu Feng, Zhi-Gui Xia, Sirenda Vong, Wei-Zhong Yang, and Shui-Sen Zhou

Operational Research Needs Toward Malaria Elimination in China
Shen-Bo Chen, Chuan Ju, Jun-Hu Chen, Bin Zheng, Fang Huang, Ning Xiao, Xia Zhou, Tambo Ernest, and Xiao-Nong Zhou

Approaches to the Evaluation of Malaria
 Elimination at County Level: Case Study
 in the Yangtze River Delta Region
*Min Zhu, Wei Ruan, Sheng-Jun Fei,
 Jian-Qiang Song, Yu Zhang, Xiao-Gang
 Mou, Qi-Chao Pan, Ling-Ling Zhang,
 Xiao-Qin Guo, Jun-Hua Xu, Tian-Ming
 Chen, Bin Zhou, Peiling Yap, Li-Nong
 Yao, and Li Cai*

Surveillance and Response Strategy in the
 Malaria Post-elimination Stage: Case
 Study of Fujian Province
*Fa-Zhu Yang, Peiling Yap, Shan-Ying Zhang,
 Han-Guo Xie, Rong Ouyang, Yao-Ying
 Lin, and Zhu-Yun Chen*

Preparation of Malaria Resurgence in China:
 Case Study of Vivax Malaria
 Re-emergence and Outbreak in
 Huang-Huai Plain in 2006
*Hong-Wei Zhang, Ying Liu, Shao-Sen Zhang,
 Bian-Li Xu, Wei-Dong Li, Ji-Hai Tang,
 Shui-Sen Zhou, and Fang Huang*

Preparedness for Malaria Resurgence in
 China: Case Study on Imported Cases in
 2000–2012
*Jun Feng, Zhi-Gui Xia, Sirenda Vong,
 Wei-Zhong Yang, Shui-Sen Zhou, and
 Ning Xiao*

Preparation for Malaria Resurgence in China:
 Approach in Risk Assessment and Rapid
 Response
*Ying-Jun Qian, Li Zhang, Zhi-Gui Xia,
 Sirenda Vong, Wei-Zhong Yang,
 Duo-Quan Wang, and Ning Xiao*

Transition from Control to Elimination:
 Impact of the 10-Year Global Fund
 Project on Malaria Control and
 Elimination in China
*Ru-Bo Wang, Qing-Feng Zhang, Bin Zheng,
 Zhi-Gui Xia, Shui-Sen Zhou, Lin-Hua
 Tang, Qi Gao, Li-Ying Wang, and
 Rong-Rong Wang*

China–Africa Cooperation Initiatives in
 Malaria Control and Elimination
*Zhi-Gui Xia, Ru-Bo Wang, Duo-Quan
 Wang, Jun Feng, Qi Zheng, Chang-Sheng
 Deng, Salim Abdulla, Ya-Yi Guan, Wei
 Ding, Jia-Wen Yao, Ying-Jun Qian,
 Andrea Bosman, Robert David Newman,
 Tambo Ernest, Michael O'leary, and
 Ning Xiao*

Volume 87

The Allee Effect and Elimination of Neglected
 Tropical Diseases: A Mathematical
 Modelling Study
*Manoj Gambhir, Brajendra K. Singh, and
 Edwin Michael*

Mathematical Modelling of Leprosy and Its
 Control
*David J. Blok, Sake J. de Vlas,
 Egil A.J. Fischer, and Jan Hendrik Richardus*

Mathematical Models of Human African
 Trypanosomiasis Epidemiology
*Kat S. Rock, Chris M. Stone,
 Ian M. Hastings, Matt J. Keeling,
 Steve J. Torr, and Nakul Chitnis*

Ecology, Evolution and Control of Chagas
 Disease: A Century of Neglected
 Modelling and a Promising Future
*Pierre Nouvellet, Zulma M. Cucunubá,
 and Sébastien Gourbière*

Mathematical Inference on Helminth Egg
 Counts in Stool and Its Applications in
 Mass Drug Administration Programmes
 to Control Soil-Transmitted
 Helminthiasis in Public Health
*Bruno Levecke, Roy M. Anderson,
 Dirk Berkvens, Johannes Charlier,
 Brecht Devleesschauwer, Niko Speybroeck,
 Jozef Vercruysse, and Stefan Van Aelst*

Modelling Lymphatic Filariasis Transmission
 and Control: Modelling Frameworks,
 Lessons Learned and Future Directions
Wilma A. Stolk, Chris Stone, and Sake J. de Vlas

Modelling the Effects of Mass Drug
 Administration on the Molecular
 Epidemiology of Schistosomes
*Poppy H.L. Lamberton, Thomas Crellen,
 James A. Cotton, and Joanne P. Webster*

Economic and Financial Evaluation of
 Neglected Tropical Diseases
*Bruce Y. Lee, Sarah M. Bartsch, and
 Katrin M. Gorham*

Volume 88

Recent Developments in Malaria Vaccinology
Benedict R. Halbroth and Simon J. Draper

PfEMP1 – A Parasite Protein Family of Key Importance in *Plasmodium falciparum* Malaria Immunity and Pathogenesis
Lars Hviid and Anja TR. Jensen

Prospects for Vector-Based Gene Silencing to Explore Immunobiological Features of *Schistosoma mansoni*
Jana Hagen, Jean-Pierre Y. Scheerlinck, Neil D. Young, Robin B. Gasser, and Bernd H. Kalinna

Chronobiology of Trematode Cercarial Emergence: from Data Recovery to Epidemiological, Ecological and Evolutionary Implications
André Théron

Strongyloidiasis with Emphasis on Human Infections and Its Different Clinical Forms
Rafael Toledo, Carla Muñoz-Antoli, and José-Guillermo Esteban

A Perspective on *Cryptosporidium* and *Giardia*, with an Emphasis on Bovines and Recent Epidemiological Findings
Harshanie Abeywardena, Aaron R. Jex, and Robin B. Gasser

CPI Antony Rowe
Eastbourne, UK
June 16, 2015